Edwin Clark

**A Visit to South America**

With notes and observations on the moral and physical features of the country, and

the incidents of the voyage

Edwin Clark

**A Visit to South America**

*With notes and observations on the moral and physical features of the country, and the incidents of the voyage*

ISBN/EAN: 9783337315023

Printed in Europe, USA, Canada, Australia, Japan

Cover: Foto ©Andreas Hilbeck / pixelio.de

More available books at **www.hansebooks.com**

A
# VISIT TO SOUTH AMERICA;

WITH

*NOTES AND OBSERVATIONS ON THE MORAL AND PHYSICAL FEATURES OF THE COUNTRY, AND THE INCIDENTS OF THE VOYAGE.*

BY

EDWIN CLARK,

*Mem. Ins. Civ. Engineers, F.R.A.S., F.M.S., etc., etc.; Author of "Britannia and Conway Tubular Bridges."*

---

LONDON:
DEAN AND SON, 160A, FLEET STREET, E.C.

# PREFACE.

HE following pages contain the substance of notes made in 1876 and 1877, during a journey to the River Plate and a residence of nearly two years in Buenos Ayres, Paraguay, and Uruguay. So far as the notes are concerned they are given in the order in which they were written, and with little alteration. They thus form a text for various observations and explanations with which they are interwoven. Such a course often involves repetition, and even discordant views on the same subject, but it has the advantage of preserving that freshness and authenticity peculiar to notes written in the presence of scenes described.

*Preface.*

I must mention that I laboured throughout under the serious disadvantage of possessing no library of reference.

The only books I possessed were Hooker and Decaizne's invaluable manual of botany, Nystrom's American pocket-book of mechanics and engineering, with its useful collection of tables and formulæ,—a remarkable compilation, but unfortunately vitiated by its many typographical and other errors; and Darwin's "Naturalist's Voyage in the *Beagle*," a book of which it is impossible to speak too highly, but which can only be thoroughly appreciated by those who have witnessed the countries he describes and realized the marvellous accuracy and eloquence of his descriptions.

Considerable attention was given to the natural history and meteorology of the country, and the notes and views on these subjects were far too voluminous for a book of this description; but though I have omitted all detailed observations, it is hoped that the meteorologist and botanist

## Preface.

will not fail to find original matter on these topics which will excite his interest.

A very large space has been devoted to the description of Paraguay, not only on account of the extreme interest which its sad history and marvellous beauty excited, but also on account of the great dearth of existing information that is available for the solution of the very remarkable physical and moral problems to which they give rise.

My professional brethren will doubtless be disappointed in finding so small an amount of information on subjects of engineering interest, but it must be borne in mind that the level plains of the pampas furnish little scope for engineering works of novelty or magnitude, while the vast and fertile regions beyond, though offering a field of unbounded importance and interest for future enterprise, are as yet unpeopled, and even unexplored. On the other hand, I have always been inclined to assign a much larger range to the subjects that essentially come within the province of the civil engineer than is usually admitted,

*Preface.*

the mere technical or practical portion forming in my opinion only a single branch of the great science of engineering philosophy, which includes the whole subject of terrestrial physics, and in that sense some of the following pages may serve as elementary lessons for the young engineer.

# TABLE OF CONTENTS.

INTRODUCTION . . . . . . . . . 9—14

### CHAPTER I.

#### THE SHIP AND THE WAVES.

Departure—Selecting a Berth—The *Douro*—Waves—Shipping a Sea—Storms at Sea—The Engines—The Slip of the Screw—The Helmsman—The Commissariat . 15—24

### CHAPTER II.

#### THE BAY OF BISCAY.

Its Extent and Formation—Its Dangers and Storms, and their Cause—Loss of the *Captain*—Ocean Depths    24—27

### CHAPTER III.

#### THE THERMOMETER ON BOARD.

Verification of—Diffusion of Heat—Indications of the Thermometer—Finding a Good Position on Board—Temperature best obtained by Rapid Motion of the Instrument through the Air—The Wet Bulb—The Whirling Table    28—30

### CHAPTER IV.

#### LISBON AND THE TRADE WINDS.

The Tagus—Lisbon—Entering the Trade Winds—Its Dread by Early Navigators—The Doldrums—Explanation of—The Atmosphere—Influence of Solar Heat—Circulation of the Atmosphere—Moisture and Heat of the Doldrums—Ascent of Heated Air—A Prairie Fire Cyclones and Anticyclones . . . . . .    31—40

# Table of Contents.

## CHAPTER V.

### LISBON TO THE TROPICS.

Teneriffe—Entering the Tropics—The Cape de Verde Islands—The *Challenger*—Coaling at St. Vincent—San Antao—Flying Fish—Nautilus—Luminous Sea—The Heat and Moisture—Absence of Lightning—Limits of the Northeast Trades and the Doldrums . . . . . 41—47

## CHAPTER VI.

### ASTRONOMICAL EVENINGS.

The Tropical Skies—The Southern Cross—The Want of an Almanac—The Celestial Equator and Meridian—The Ecliptic Defined by the Moon—Fig. 1.—Southing, Rising, and Setting of Planets—Jupiter's Satellites and Longitude—The Celestial Globe—The Compass and its Corrections—The Log—The Latitude, Longitude, and Sun's Azimuth . . . . . . . . . 48—58

## CHAPTER VII.

### VAPOUR AND CLOUDS.

Specific Properties of Water in all its Forms—Its Principal Functions—Its Anomalous Expansion—Tropical evaporation—Tension of Vapour—Increase of Volume—Diffusion of Vapour—Two Atmospheres—Latent Heat—Mechanical Force of Evaporation — Lightness of Saturated Air—Specific Heat of Water—Table of Properties of Vapour—Its Influence—Daniell's Summary—The Dew Point and Table for finding . . . 59—67

## CHAPTER VIII.

### ST. VINCENT TO THE PLATE.

Crossing the Line—San Fernando—Globigerina—The Southeast Trades—Pernambuco—Tropical Flora—Bahia—The San Francisco Line—Luxurious Vegetation—The Ants—

# Table of Contents.

Cape Frio—Sea Temperature—The Gulf Stream—Rio—Mountain Clouds—Botanic Gardens—Tejuca—The Yellow Fever—Fruits—Climate—A Shower—Fig. 2.—The Terrestrial Layer—Change of Climate—The River Plate—Yellow Fever on Board—Quarantine Regulations—Monte Video—Anchorage at Buenos Ayres . . . . . 68—92

## CHAPTER IX.

### QUARANTINE AND THE RIVER PLATE.

Difficulties—Death on Board, Yellow Fever—Funeral—A Pampero—Fig. 3.—The Quarantine Hulk—Cruel treatment—Description of the Plate—Silting—Delta of the Parana—Great River Valley—The Baranca—The Las Palmas and Guaza Branches—Campana—The Tides—Theory of, and Table—Amusements—Humidity and Dew—A Cloud Observation and Storm—The Camalote and Botany—The Islands—Air Plants—The North Winds—Our Acclimatization—New Arrivals—Picnics—River Fogs and Cloud Formation—A Distant Storm—Sail to Buenos Ayres 92—123

## CHAPTER X.

### BUENOS AYRES AND THE PAMPAS.

First Impressions—Narrow Streets—Bad Drainage—Bad Roads—Buildings—Italian Workmen—Religious Freedom—The Harbour and Lighters—The Campana Railway—The Animals and Birds—The Marshes—A Tumulus—The Baranca—Thistles—The Great River—Progress of Civilization—Engineer's Difficulties—The Northern Railway—The Ensenada Railway—Forest Planting—The Pampas—Its Gradual Rise—Its Formation—Absence of Trees—Its Natural History—The Southern Railway—No Fences—Azul, a Night at Dolores—Ephemeral Nature of Civilization—Floods of the Salado—The Western Railway—Its Benefits—The Crisis at Buenos Ayres—Public Loans—Great Cost of the Government—Defects—The Public Works—Maladministration—Hospitality of the people . . . . . . . . . 124—162

# Table of Contents.

## CHAPTER XI.
### CLIMATE AND METEOROLOGY OF BUENOS AYRES.

Advantages of its Peculiar Position—Freedom from Local Disturbance—The Pampas—Constant contest between Polar and Equatorial Currents—Periodicity and Cause of Storms—The Terrestrial Layer—Overflow of Equatorial Vapour—Rapid Condensation—The Andes Storms Travel against the Wind—Fig. 4—The Phenomena of a Great Storm—Its Analogy with Condensed Steam in an Engine—The Great Storm of May 1877—Barometer Observations—Figs. 5, 6—Pressure from Descending rain—The Lightning—The Floods—The Damage to the Railway—Restoration—Fig. 6a—Conclusions—Storm of June—Its Extent—The Santa Rosa storms of August—Lightning, Duration and Extent—A Dust Tornado—Vapour Overflows from the Tropics in Rolling Cylinders—Fine Climate of Buenos Ayres—Coincidence of Storms and Full Moons—Sunsets . . . . . . . . . . 163—204

## CHAPTER XII.
### THE URUGUAY AND SALTO.

Magnitude—Estuary and Delta of the Uruguay—The River Steamers—Paysandu—Fray Bentos—Salto—Cause of its Decline—Public Loans—Dishonest Government—The Present Dictator—Recent Improvement and Progress—The Salto Railway—The East Argentine Railway—Description of the Salto line . . . . . 205—216

## CHAPTER XIII.
### THE COTTAGE AT ARAPEY.

The Bridge—Solitary but Charming Position of the Cottage—The Commissariat—The Staff—The Wind Vane—Fig. 7—The Azimuth Table—The River—The Monte, Alligators—Stray Cattle on the Monte—Charms of Solitude—The Natural History—Botany of the Rocks and Monte—The Lagoons—A Striking Example of Natural Selection 217—232

# Table of Contents.

## CHAPTER XIV.
### NOTES AT ARAPEY.

The Locusts—Callandrias—Widah Birds—A Puma—Lightning—Powder Magazine—The Camadraca—The Iguana—The Cuati—The Nutria—The Carpincho—Rodents—The Snakes—The Lakes—Alligators—Scorpions and Spiders—The Becho Colorado—Puncture of the Mosquito  233—244

## CHAPTER XV.
### METEOROLOGY AT ARAPEY.

The Heated Rocks—The Heat and the Flowers—Fire Flies and Night Insects—Effects of Heat—Experiments on Radiation—Ascent of Heated Air—Mean Temperature and Weather—Dust Storm—Tornadoes—Explanation of—Resemblance to a Centrifugal Pump—Their Movement—Causes of Storms—Burning the Flechilla Grass—Clouds—Their Shadows—Their Formation—Perspective of—Height and Forms of—Temperature of—Length of Shadows as Affecting Climate—Excessive Heat and Drought at Christmas . . . . . . . . 245—263

## CHAPTER XVI.
### PARAGUAY AND ITS HISTORY.

Its Beauty and Sad Fate—Its Destruction—The Guarani Indians—The Jesuits—Rapid Development of the Country—Recall of the Jesuits—Dr. Francia and his Cruelty—Lopez I.—Prosperity under His Rule—Lopez II.—His Bloodthirsty Disposition and Vices—State of the Country at his Death—The Railway—The Loan—The Tobacco Confiscation—President Gill's Assassination—Projected Union with the Argentine Confederation . . 264—274

## CHAPTER XVII.
### THE PARANA AND ASUNCION.

Magnitude of the Parana—Its Tributaries—Field for Exploration—Distances—Temperature of Water—The Baranca of the Parana—Geology—The Gran Chaco—The Indians—Their Degeneracy—Excelsior — Change of—

Scenery—Shooting Monkeys and Alligators—Corrientes—
Asuncion—The Lagoons—The Victoria Regia—The Chaco
—Description of Asuncion—Francia's Palace—Lopez's
New Palace—Vast Resources of the Country—Our Hotel—
Undue Proportion of Females—The Market—The Railway
—The Environs and Natives—Sale of the Railway   275—296

## CHAPTER XVIII.

### THE COTTAGE AT PARAGUARI.

Our Landlord—The Village—Contrast with the City—Our Fare
—Amiable Character of the People—The Commandante—
His Residence—The Schools—Absence of Religious Culture
—Municipal Accounts—Power of the Commandante—
Absence of Crime—Incident at a Wedding—The Market
Produce—The Village Dial—Eclipses—The German
Colony—The Forest Scenery—Native Houses and Habits
—Orange Gardens and Banana Groves—Bamboos—
Botany—Hospitality—Sugar and Mandioca—Snakes—
Canna Distillery—Sporting—The Orange Distillery—The
Boa-constrictor—St. John's Cave—Balls—Our own ball—
Peaceful Character of the People—Physical Cause of the
Great Fertility—Climate—Siesta—Meditations on Happiness—Theory of the Black and White Balls—Mosquitoes
—Jiggers—Amusements and Departure—The Great Heat
at Asuncion—The River Journey Home—The Oranges at
Ipane—Our Cargo—Oranges at Pillar—Running Aground
—Our Delays and Hardships—Arrival at Rosario   297—336

## CHAPTER XIX.

### CORDOVA AND THE SIERRAS.

The Railway to Cordova—The Rise of the Pampas—The Barren
Country—Situation of Cordova—The Jesuits—A Procession—The Alameda, or Artificial Lake—The River—The
Tucuman Railway—Our Scientific Friends—The Astronomer, Mr. Gould—The University Professors—Botanizing—Expedition to the Sierras—Picnic—The Mill on
the Tercera—The Lasso—The Botany—The Observatory
—Selection of its Site—The Sierras—Return Home 337—347

APPENDIX . . . . . . . . 349—355

## INTRODUCTION.

IT is a very natural desire with every intelligent traveller to commemorate by some record the privilege he has enjoyed, the experience which he feels he has acquired, and the prejudices which he knows he has overcome.

The first aspect of nature and natural phenomena in all the imposing grandeur with which she surrounds her tropical throne, affords to the philosophic mind an intensity of enjoyment totally indescribable, and we too often fall into the error of imagining that we can interest others by a bare record of our own delight; but it is equally impossible either with pen or pencil to convey any adequate idea either of the magnificent luxuriance of a tropical forest, or the terrific grandeur of a tropical storm. Such phenomena are indescribable, and must be witnessed before they can be imagined.

One is thus reduced to the dry and unsatisfactory task of furnishing a barren list or description of the plants of which the forest is composed, or the registered force of the wind or temperature or rainfall of the storm.

This has not been the dominant object in the following notes. The enjoyment of such scenes is greatly dependent on our preparation for their comprehension, and is vastly enhanced by a sufficient knowledge of natural philosophy and natural history to look on them not only as gorgeous and imposing spectacles, but as objects of scientific interest. We are irresistibly led to inquire why all the familiar objects of our everyday life, put on such strange and novel forms beneath the mysterious influence of this heated zone ; why our hills and valleys here are carpeted with such modest vegetation, while gigantic productions of totally new types and forms render these prolific regions literally inaccessible from their very luxuriance, covering the land with palms and cycads, gigantic grasses and ferns, or with forests in which the trees themselves are with difficulty identified amid the cloud of epiphytes and climbing plants and parasitic vegetation which covers and overtowers them.

In a similar manner we attempt in vain to utilize the experience acquired in our gentle climate, and apply our weather theories to these regions of vapour and heat, where all meteorological phenomena assume such gigantic proportions, and, indeed, where all storms originate.

It is in the presence of scenes so novel and so imposing that we feel so deeply not only our own deficiencies and want of preparation for such an intellectual feast, but the infinitesimal smallness of human knowledge altogether.

One striking lesson derived from travelling is the limited field to which all observation is rigidly confined. The smallness of the earth is prominently brought before us when we reflect that the Royal Mail Steamers make a quarter of the circuit, with a regular fortnightly service, for a few pounds sterling, between here and the Plate. The journey occupies twenty-eight days.

Again, small as it is, three-fifths of the surface of the earth is a mysterious reservoir of water, and the whole is covered by so scanty an envelope of atmosphere that an elevation of six or seven miles, which can be performed in a balloon in half an hour, would instantly annihilate the whole human race, while a similar descent beneath the unstable, cooled, superficial film on which we live would be equally destructive. The visible abyss of space around us reflects our insignificance, and, vast as it appears to our limited conceptions, it is doubtless only another film as compared with the inconceivable infinity beyond.

Some four years since an intelligent country apothecary, who knew my fondness for natural history, brought me an ordinary wide-mouthed chemist's bottle, containing, as an article of commerce, dried, pounded hemlock leaves, labelled, "Fol, Conii." The bottle was more than three-fourths full. The bottom layer, to the extent of half the bottle, consisted of the original dried leaves, but the top layer, about 1½ inches thick, consisted of a bed of matter of a darker colour, the exuviæ and remains of former generations of a colony of thousands of minute coleoptera who lived on the surface of the sound portion of the dried leaves.

He informed me that this upper bed, the cemetery of former races, had been ten years in course of formation, fresh colonies being periodically produced. The bottle, during the whole of this long period, was kept constantly closely corked. In this state he gave it me four years ago. It has been on my mantle-piece ever since, and is now before me with this year's progeny of beetles visible to the naked eye, moving about in a dense layer over the remaining material, which is still sufficient in quantity to support many future generations in this miniature world.

That these beetles should thrive so many years in a closely-corked bottle of leaves of so poisonous a plant as hemlock is of itself a remarkable instance of adaptation, but I have not recorded it on account solely of its entomological interest.

I have often thought that if these beetles were endowed with intellect of the highest order, and brain power capable of indefinite development, how infinitely small must be the ultimate sphere to which all their philosophy must be limited; and I have asked myself whether our own bottle is much larger as compared with the universe in which it forms so insignificant an atom.

They would doubtless see through the glass that forms the confines of their tiny world the rows of other similar bottles which occupy the chemist's shelves, and would find them in so many respects like their own that they would doubtless conclude that they must all be full of beetles, in the same way and with the same probability that our speculative astronomer peoples Jupiter and Saturn with beings like ourselves.

Night and day, even summer and winter, must be familiar to them, and would doubtless be explained by innumerable theories without the slightest possibility of their arriving at any correct explanation. May this not also be the case with some portion of the knowledge of which we are vainly in search?

Their only geology, very much resembling ours, a bed of exuviæ and skeletons of their fossilized ancestors, would show them that previous races, more or less highly developed, had run the same course as themselves under similar conditions, and the deep, sound bed still remaining beneath them would, as regards their scale of time, ensure an infinite future.

I am often, as I have said before, inclined to ask myself whether in the great scale of creation our little globe and its boasted geological antiquity and assumed eternity and stability may not after all be regarded only as a little bigger bottle in a larger chemist's shop.

It is in this spirit that I have in all humility attempted to utilize the small amount of natural philosophy which it has been my lot to acquire; first, in endeavouring to explain the natural laws under which some of the phenomena I have witnessed range themselves, and, secondly, like the beetles in their speculations on the seasons, to theorize in the best way I can on the much greater amount of phenomena of which we know nothing.

## CHAPTER I.

### *THE SHIP AND THE WAVES.*

WE left Southampton on the 10th April, 1876, bound for Buenos Ayres. The *Douro*, one of the older vessels of the Royal Mail Steam-Packet Company, in which we sailed, was an excellent sea-boat, with accommodation which left nothing to be desired. We went on board, in good English fashion, during a storm of rain with a strong cold westerly wind, which lasted for three days, all across the Bay of Biscay, till we sighted Cape Vincent. We went through the usual discussion as to whether berths should be transverse to the ship or longitudinal, and we had sufficient pitching and rolling to add zest to the arguments. In this case, as in the majority of cases in which disputants disagree, there would have been no difference of opinion if we had all started with the same premises. No one likes their head alternately a foot above and below their heels, and if vessels only rolled we should certainly be most comfortable parallel with the keel; but when steaming against a head wind, it is often not the rolling but the pitching that bewilders our ideas of gravity, and upon the whole I came to the conclusion that to be thoroughly comfortable we ought to sleep in hammocks, suspended in gimbals, like the ship's compass, which always preserves such provoking and enviable stability. But even with these elements of horizontality we still have the rise and fall due to the mean height of the great water undulations. It is hopeless to attempt to neutralize such complex and ceaseless movements by any mechanical device, and a journey across the Bay effectually removes all illusions on this subject.

It may be a useful hint to a landsman selecting his berth, to remind him that if he stands at the stern of the vessel looking ahead, his left hand is the larboard or port side, the letter L being the initial of both left and larboard, and his right hand is the starboard side, and as the course across the Atlantic to South America is about south-west, and the sun at starting is south, he will prefer, if possible, a berth on the starboard side, as very much cooler than on the port side, and *vice versa* on his return.

On a first ramble over a fine vessel like the

*Douro*, we cannot fail to be struck with its strength and solidity, and the numerous and ingenious appliances involved in its construction. As we pace the spacious and solid deck we feel elated with such a triumph of human skill and industry, and half inclined to defy the winds and the waves. The Bay of Biscay in its angry moods is a good school for checking all such presumption.

It must be borne in mind that an ordinary wave in still water is a mere undulation in which the particles of water that compose it do not move forward, as they appear to do, but oscillate in vertical lines up and down. Hence an object such as the vessel floating on such a wave is not carried forward. Such an undulation requires a certain depth of water, and no great waves can be formed in shallow water. Where the wave meets a vertical wall, it simply rises up and down on the face of the wall, but if it meets a shelving shore its vertical oscillation is arrested from want of depth, and the force is expended by its conversion into a wave of translation, which flows bodily on to the beach, carrying objects rapidly forward; the retiring water runs back beneath the advancing wave, which, on account of its forward velocity, curls over and expends itself in incessant breakers that follow each other with rhythmic regularity.

A similar undulation is the base and essence of

all waves, and of the billows of broken water that characterize the Bay of Biscay. An instantaneous hurricane on still water would merely bear away a thin layer, into spray; but when the wind impinges on the sloping side of these undulations it is resolved into two forces, one tending to depress the weather side of the wave, or practically increase its weight, and thus increase the height of the succeeding wave into which it resolves itself; the other force tending to carry the undulation bodily forward and produce a wave of translation, such as we have seen breaking on the beach. In a similar way while friction delays the lower particles, the top layer of the wave hurries furiously on, and curls over with foam and noise under the action of long-continued wind, which always blows in gusts.

The oscillations of this liquid pendulum attain colossal dimensions, although even during a gale the waves maintain a definite form and a partial regularity in their movements. Under such circumstances the good ship rides over them, pitching and dipping into successive hollows with heavy shocks, when facing the storm, and shipping incessant spray and sometimes solid water. When running before the gale her motion is easier, but a huge, permanent, overhanging wave appears to follow in her wake, rivalling her own speed, and threatening to overwhelm her,—a catastrophe which does occasionally occur. But if through the breakdown of the engine, and want of way or damage to the rudder, the vessel no longer obeys her helm, but drifts helplessly into the trough of the sea, she then rolls heavily from side to side, dipping her bulwarks into the sea, while huge billows break from stem to stern, clearing the decks of all unsecured obstacles, streaming down the hatchways, and deluging the engines and furnaces, and sometimes adding to the helplessness of the crew by extinguishing the fires.

But the Bay must be seen during the fury of a hurricane before any idea can be formed of the prodigious forces with which a vessel has to contend. On such occasions the billows rival the winds in their wild and lawless fury. Mountains of water clash together or combine, and are converted into whirlpools and foci in lines of irresistible force. Immense masses of water scud away before the tempest with velocity almost equal to that of the wind itself, often drifting directly across, or even in direct opposition to, the general set of the waves. Such abnormal disturbances are fortunately few and far between, and the helmsman endeavours to avoid them as best he can; but when such a moving column, or "sea," as it is termed, strikes the unfortunate vessel, its effects are terrible and destructive beyond explanation. Solid iron stanchions are bent flat down on the deck, bulwarks, boats, and every projection, often including the hatches themselves, with every unfortunate being that comes within its range, are swept bodily away. The hull trembles under the blow as though striking a rock. The course of the vessel is instantly arrested, and it only too often happens that she rises out of the sea again a helpless wreck.

In these large steamers sails are but a secondary resource. The great breathing engines are the true life of the ships, and are nursed during a storm with scrupulous anxiety and care. They are only driven at slow speed, but it continually happens that during the violent lurches of the vessel her stern and screw are lifted entirely out of the water; no governing arrangement can act with sufficient speed to cut off the steam, and the tremendous force of the liberated engine is for the moment totally expended in imparting terrific velocity to the shaft and screw, not unfrequently damaging the engines or twisting the great shaft asunder. This is guarded against as far as possible by an engineer in charge of the throttle valve, who watches as well as he is able the pitching of the vessel, and temporarily cuts off the steam by hand at the moment he has reason to apprehend the stern is leaving the water. But in spite of these precautions the screw occasionally will run away, and every traveller must have been aroused by the startling jar and tremor which in such cases vibrates through the whole vessel from stem to stern.

To guard against accidents at the helm, vessels are generally provided with duplicate steering gear, the noisy chain and quadrant tiller which is worked from the bridge, and the right and left handed screw-gear worked by the ordinary ornamental steering wheel at the stern. Whoever watches the steersman at these wheels during a gale must be struck with the dangers of their arduous duty, and must regret that safety is not preferred to ornament in so vital a part of a ship's equipment. It is difficult to imagine a more dangerous position than that of a sailor on an inconvenient raised platform in a rolling ship, between these two wheels with their projecting spokes. If a blow of the sea overcomes them, the wheels rotate with terrible velocity, and seem constructed as traps for arms and legs. It would perhaps be better if the wheels were plane discs or protected by discs, but the shaft should always be furnished with efficient elastic brake power, with a long lever and proper locking gear for use during bad weather. The steam steering machinery now coming into use must be a great boon. Hydraulic machinery is also applicable to this purpose.

It requires a few days to realize the extent of the immense establishment maintained on board these large vessels. The officers, crew, and servants number 115 people. We had in addition, including those taken on board at Lisbon, 130 passengers. The commissariat department, considering the many claims on the ship's space, is provided in no niggardly scale. The fore-deck is crowded with cattle-pens, and cages filled with oxen, sheep, pigs, and poultry, affording abundant fresh meat throughout the journey. Ocean-acclimatized cows supply fresh milk, and the ship is provided with ice cellars, with forty tons of ice,—a luxury which is liberally dealt out, and thoroughly appreciated, throughout the tropics. A large tank placed in the centre of the ice-house supplies the table *ad libitum* with filtered iced water, while a well-equipped bar supplies iced American drinks of all kinds. Some kind passenger generally presides at the piano in the great saloon, and there is no difficulty in joining a whist party on a level with one's own skill and experience.

Altogether we were a very happy family, and although the sea always imparts a certain amount of its own restlessness and prevents close application, it affords abundant scope for observation, more especially as regards its own mysterious phenomena and the constantly-varying aspect of that great and marvellous vault that bounds the unobstructed horizon with all its gorgeous furniture of clouds and mists, dimly veiling the moving hosts of stars and planets beyond.

## CHAPTER II.

### *THE BAY OF BISCAY.*

WE crossed the Bay of Biscay from Ushant to Cape Finisterre, a distance of about 425 miles, in about two days, or forty-eight hours, with a strong cold westerly wind. This enormous basin extends inland to Bordeaux, a distance of 300 miles. Its southern boundary runs due east and west along the parallel of 43·30 with singular regularity. This inhospitable coast, backed by the mountains of Asturia, consists of a wall of silurian and crystalline rocks,—a continuation of the great Pyrenean formation, with carboniferous and Devonian measures, extensively worked in the well-known mining districts of Santander and Bilbao.

Its northern boundary consists similarly of an irregular, deeply-indented coast, also composed of hard crystalline metamorphic rocks, and the formation of the Bay appears partly due to the action of the great Atlantic wave, confined between these indestructible walls, thus clearing away the tertiary measures which line the bottom of the Bay from Rocheforte to Bayonne. A continuous denudation of this character would ultimately absorb the whole district of the Garonne and form a new entrance into the Mediterranean, flanked by the Pyrenees. During westerly winds, which set directly into the Bay, no sailor after losing sight of the great lighthouse at Ushant is free from anxiety until he sights the flashing beacon off Cape Finisterre, for there is little chance of escape for a crippled vessel driven into this stormy *cul-de-sac* during a continuous westerly gale. The Bay owes its dangerous character not only to its land-locked, inhospitable coast, but also to the strong and shifting tidal currents which set around the Northern Cape, and to the fact that the great Atlantic waves assume unusual dimensions and irregularity from their reflection as it were into foci by the concave coast. It is moreover peculiarly subject to violent storms, from the circumstance that it is situated in the boundary which separates the great zone of steady trade-winds from the region of cyclones and disturbances which characterize the North Atlantic. The numberless and disastrous wrecks which from time immemorial have given such unenviable notoriety to these stormy waters, impart a melancholy interest to its history, crowned by that gravest of all marine disasters—the loss of H.M. turreted vessel the *Captain*, with 500 men on board, of whom only eight or ten escaped to record the mysterious lessons of the disaster.

The great depth of the water is also a remarkable feature, amounting to 2600 fathoms within the Bay, so that we crossed no deeper water in the whole of our journey to the Plate. The general average depth of the Atlantic is 2000 fathoms, or about 2¼ miles: the height of Mont Blanc, as a convenient scale of reference, is 2500 fathoms. The depth, however, in the great depression south of the Bermudas amounts to 3875 fathoms, which is the greatest known depth in the Atlantic Ocean. See the cruise of the *Challenger*.

## CHAPTER III.

### THE THERMOMETER ON BOARD.

WE made daily observation of the temperature of the surface water, and also of the atmosphere. The instruments were carefully verified by comparison with a standard, and tested in a tin vessel filled with pounded ice and rain water, and their errors at freezing point were ascertained. The best instruments are liable to slight changes of their zero, from contraction of their glass bulbs and stems, more especially when exposed to sudden changes of temperature. These repeated tests are useful lessons as regards the accuracy of the instruments themselves, and the care and attention required in their use.

The problem of finding the best position for the thermometers gave rise to a few experiments. Heat is communicated to a thermometer either by direct contact with other bodies, as when immersed in water or exposed to a current of air; or, secondly, by radiation or reflection from other objects. Heat radiates in all directions from the thermometer, and from surrounding objects, like light from a candle, until an equilibrium is obtained by which the heat received and emitted becomes equal. A free thermometer never absolutely attains, but indefinitely approximates to and records the general mean effect of numerous surrounding causes. It is affected by reflection, and still more rapidly by direct contact with other bodies, such as the air itself; but the temperature of the air is only one among other influences. Its actual indication is dependent not only on the movement or tranquillity of the air, but on the nature, distance, and even superficial coating of surrounding objects.

When we consider that different bodies have different capacities for heat, different powers of radiating, absorbing, and reflecting heat, and different capacities for transmitting and conducting heat, it is evident that the combined influence of all these properties of heat on a glass bulb filled with mercury only furnishes us with a record of the mixed effect of numberless disturbances, and not a simple record of the temperature of the air, which is the real object of our inquiry. Thermometers placed in different parts of the ship all gave discordant results: one gave the temperature of the warm walls of the cabin; another, situated outside a port-hole on the port side, gave us the temperature of the air, compounded with that of the warmed side of the iron ship, and of the reflection and radiation from the sea water. On the starboard side we had a lower result, because that side of the iron vessel was in constant shade. On the main deck we found that the heat which came from the furnace and funnels raised a thermometer constantly two degrees. The bows gave too low a result, because the spray exposed them to the effects of evaporation. Any height up the mast was objectionable on the score of inconvenience, and the difficulty of shading the thermometer from the direct rays of the sun. In fact, a good situation to-day became a bad one to-morrow, and although a permanent set of thermometers was placed on the shady side of the after hatchway, properly protected by Venetian shutters, I preferred to adopt at the time of observation the best temporary position available for the object in view. For this purpose I availed myself of the rapid conduction produced by rapid motion through the air, by which other influences are dwarfed, and I used a delicate thermometer attached on a sling, and whirled rapidly round in the air in a properly selected and shaded situation. The thermometer being first warmer than the air, is whirled round until it ceases to fall, and this is recorded as the exact temperature of the dry bulb. I then wrap the bulb of the same thermometer in a strip of wetted lint, and whirl it again until it again ceases to fall; this result is the temperature of the wet bulb. On removing the lint and drying the bulb it is again whirled, and its exact return to the previous temperature is a proof of the accuracy of the experiment, which scarcely occupies two minutes. When on shore I fixed it on a whirling table, with a proper protection against the heat of the body by means of a glass cover over a portion of the table, through which the thermometer is read by means of a lens, and I found all the results thus obtained highly satisfactory and uniform.

## CHAPTER IV.

### *LISBON AND THE TRADE WINDS.*

THUS constantly occupied, the days passed happily by, and the ship's bell, that called us to incessant meals, always disturbed some interesting investigation. It was a constant practice, even during the night, to come on deck once or twice in slippers and dressing gown, to have a look at the weather, and a few minutes' meditation on the changing aspect of the star-lit vault above us. The cold, wet westerly Channel winds, with their accompanying fogs, left us, or rather, we left them, by the time we rounded Cape St. Vincent, and on the following day, Good Friday, April 14th, 1876, after a four days' journey, we anchored in the Tagus at Lisbon. One cannot fail to be struck with the elegant proportions and chaste decorations of the old castle of Belem. What is the reason that the old architectural monuments, whether forts or churches, are invariably so much more imposing than anything we seem now capable of designing? We landed for a few hours under a brilliant sun, and took a drive round the Wind-Mill Hill. The glimpses we got of large aloes, and arums, and palms, and other outscouts of tropical vegetation, excited our enthusiasm. We took on board a boat-load of oranges and other fruits, with cattle and fish and vegetables, and about 20 first and 120 third class passengers. The news that yellow fever was unusually prevalent in Rio, and dread of the proverbial miseries of quarantine in the Plate, deterred many persons from joining us who would otherwise have done so. Before we had made much progress in our return down the Tagus, we were met by a strong N.E. wind, amounting at first to half a gale, which lay on our starboard bow, and rolled us about sadly all night. On leaving the river, however, and resuming our normal S.W. course, we ran merrily before it until it moderated down to a pleasant steady breeze. This same N.E. wind accompanied us without variation for nine days and nights, through a distance of about 2400 miles, until it assumed a more easterly direction, and died away imperceptibly amid the heated clouds and vapours of the Doldrums in Lat. 5 north. This was our first acquaintance with the north-east trade winds, which temper these quiet climes. The marvellous constancy of this eternal wind might well alarm the early navigators, who after running before it day after day, despaired of a change and feared they could never return home. But the zone of calms at the Doldrums in which it landed them was far more to be dreaded. In this region of breathless heat and vapour, sailing vessels are often helplessly becalmed for weeks together. So motionless is the air that the smoke of a cigar scarcely leaves the deck, and the damp sails hang useless among the rigging. There is no exaggeration in Coleridge's beautiful metaphor, in describing this lifeless scene :—

> " Down dropt the breeze, the sails dropt down :
> 'Twas sad as sad could be ;
> And we did speak only to break
> The silence of the sea.
>
> " All in a hot and copper sky
> The bloody sun, at noon,
> Right up above the mast did stand,
> No bigger than the moon.
>
> " Day after day, day after day
> We stuck, nor breath nor motion,
> As idle as a painted ship
> Upon a painted ocean."

The explanation of this singular phenomenon offers however no difficulty, though it leads us into a little dry philosophizing on the general subject of the winds.

The atmosphere above us, like the ocean beneath us, though homogeneous as regards actual composition, is not so as regards its density. If we ascend into the air, the density of the atmosphere constantly decreases in so rapid a proportion that at a height of only $3\frac{1}{2}$ miles it is already halved, and its temperature is proportionately diminished. The winds with which we are concerned are the disturbances of this lower envelope of dense air. This unstable terrestrial stratum, constantly varying in density and temperature, must be subject to perpetual changes and movements. And after our experience of the storms to which the surface of so dense a medium as the ocean is liable, we are fully prepared for the corresponding hurricanes and tempests of this rarer medium. In the same manner as we have been able to explain the great dominant features of an ocean wave, without being able to give any satisfactory account of the abnormal and destructive forms which it assumes during a hurricane, so also, although we are able to trace the great fundamental movements of the atmosphere, we are often unable to unravel its lesser and more complicated phenomena. The tempests which roll the Atlantic billows into the Bay of Biscay, and slowly disintegrate a continent, are due solely to the energy

## The Atmosphere. 35

of solar heat. In every daily circuit a wave of creation and destruction accompanies this mysterious orb in its journey round the earth. Its universal influence is the basis of almost every physical problem, but is nowhere more directly manifested than in the phenomena of the winds and waves. The lower stratum of the atmosphere, however, owes its condition more immediately to terrestrial influence. Its temperature and humidity are mainly due to the surfaces with which it comes into direct contact. Solar heat radiates freely through the air and reaches the earth with little loss. That portion which falls on the land is rapidly absorbed and retained, while that portion which falls on the ocean is only partly absorbed, a large proportion being consumed by its conversion into latent heat, while in both cases a portion is reflected back into space. Apart therefore from all minor disturbances, the earth and atmosphere as a whole are much more highly heated in the neighbourhood of the equator, where the sun is vertical, than under the influence of its oblique rays in northern and southern latitudes; continents and islands moreover are relatively more heated than oceans and seas. These local differences of temperature give rise to the trade wind.

The void occasioned by the constant ascent of columns of heated air from the belt between the tropics is filled in by a constant current along the surface, and if the earth were at rest would give rise to constant winds from north and south.

But when we bear in mind that the equatorial belt and the atmosphere with it rotates with a velocity of 1000 miles per hour, while there is no such motion of rotation at the poles, it is evident that if a cannon-ball could be instantaneously transposed from the pole to the equator it would from its inertia stand still; while the earth whirled past it from west to east, it would thus be endowed with an apparent motion of 1000 miles per hour from east to west. This same effect would be produced in a proportionate degree by any change of latitude, in the ratio of the cosine of the latitude, and with any other heavy body such as the air. In this manner winds coming direct from the north arrive at the equator as north-east or even east winds; while winds from the south pole apparently come from south-east. Apart from other disturbing causes there is thus a general tendency to a circulation of the whole atmosphere in which cold winds flow in along the surface of the earth towards the tropics from north-east and from southeast, while the heated air from the tropics rising into the atmosphere flows away north and south to restore the equilibrium.

It is moreover evident that the ascending columns of heated air from the whole of the equatorial zone would on this hypothesis become compressed as they approach the pole, and attain parallels of latitude of less and less diameter, as shown by the approach to each other of the meridians on a globe. The effect of this compression is to bulge the atmosphere as it were upwards and increase its height, and this bulging effect becomes very apparent and constant in about Lat. 32° north, where it is characterized by a permanently high reading of the barometer. Between the equator and latitude 32°, the constancy of a circulation of this character is very evident, although it is further disturbed by the difference of temperature between the land and the sea.

The great belt from which these columns of heated vapour rise varies in position with the sun. At sea it is called the Doldrums, and is characterized by its moisture, its high temperature, and its copious showers, which, near the equator, are incessant. In the movement of these currents over the surface, in their descent as well as in their ascent, and in the work done by evaporation, enormous mechanical energy is consumed, which is wholly supplied by a constantly vertical sun. In the form of clouds, and storms, and rains, these labours of the sun are spread from pole to pole.

Direct mechanical movements of the air or winds are produced by the void left by condensation of vapour into rain, or by the descent of cold winds from higher levels or mountains, or by the local ascent of heated or humid columns of air, on account of their diminished specific gravity. The great quiet trade-winds arise from the heating of a whole zone of atmosphere, while the land breezes and local storms are produced by the superheating of the continents as compared with the ocean.

If we merely light a bonfire, or, as we should do in South America, set fire to a forest or prairie, we witness on a small scale the elements of a great storm. Columns of flame and smoke rise furiously into the heavens, and the ascending current draws in the air in violent eddies from every side to fill the void. The barometer indicates the diminished pressure, and rain is often immediately produced by descending currents of cold air. This same phenomenon takes place when one part of a continent consisting of darker rock or sand is more highly heated than the surrounding districts covered with marsh or forest, or when clouds partially obscure the continent, leaving large tracts exposed to full sunshine, or when a cold polar current overflows irregularly the heated pampas, or a heated equatorial current penetrates a colder zone, and in numberless other similar circumstances.

## Cause of Winds. 39

It is interesting to examine how such a void will be filled. First, the air will flow in from all sides, but principally from that side on which the actual barometric pressure is greatest, and with a velocity proportionate to the gradient of pressure. This gradient is indicated graphically by a series of isobars drawn at equal barometric intervals round this area of lowest pressure. Secondly, in the northern hemisphere the air flowing from the north will, as we have seen, on account of the earth's motion, come in from the north-east. Similarly the air from the equator will come from the south-west, and thus all round this region of low pressure the winds, instead of flowing directly towards it, will tend to rotate round it spirally.

We have therefore this remarkable and important general law, that all winds rotate round a region of low pressure in a constant direction, viz., in northern latitudes in a direction contrary to the hands of a watch. Such systems of winds are called cyclones. A succession of such cyclones, which continually pass over us, is a main cause of our unstable weather in England. They are often of great extent and violence, and come to us generally from the west or south-west, often originating on the northern boundary of the region of the trade winds, with which we are more immediately concerned. Winds which blow around a region of high pressure, or anti-cyclones, rotate necessarily in a contrary direction, viz., with us in the same direction as the hands of a watch.

Thus winds never blow in straight lines, but in curves, and in England, if we face the wind during a cyclone we know the centre round which it is blowing, or the region of low pressure, must always be on our right hand.* This important and constant law is the basis of all our weather forecasts, but as we approach the equator it is totally merged in that higher law of circulation which, as we have seen, is more directly applicable to the whole region of the trade winds.

## CHAPTER V.

### *LISBON TO THE TROPICS.*

T was after leaving Lisbon in latitude 38° 30′ that we experienced a marked change of climate, and felt that we were entering a new meteorological region, the great province of the sun, which we did not again leave for nearly two years. Accompanied by our new friend the north-east trade wind, and its usual escort, a little mist and cloud, we ran about 260 miles a day, with a consumption of about fifty tons of coal. We passed the Peak of Teneriffe in the dark, but in the morning light we caught a glimpse of its stupendous dome among the clouds, 12,180 feet above the sea. It was clad with snow. Its great height is available as a valuable measure of the altitude of clouds and currents; at its summit the wind is usually directly opposite to that of the constant trade-winds below, and consequently we see the over-

flow from the equator passing back over our heads. On the 18th of April we entered the tropics, latitude 23° 30′, and enjoyed the first truly tropical day with a hot sun, now nearly vertical at noon, although masses of fleecy cloud with haze still accompanied the north-east breeze, and tempered the nights. On the 20th we arrived at that singular volcanic group the Cape de Verde Islands, and rounding San Antonio anchored in the deep and brilliant and sheltered waters of St. Vincent. The *Challenger*, on her return voyage, was anchored in the bay alongside us, and we were most courteously invited to visit this now weather-beaten and interesting floating museum, and inspect her perfect scientific equipment. This little break was most enjoyable and opportune. The temperature in the cabins was 82°, and on deck 85°, the water was 66°, and, to add to our discomfort, the whole of the ports and hatchways had to be tightly closed, as we had put in here for the purpose of undergoing the terrible operation of coaling.

The two most serious nuisances in a steamer are doubtless the coaling of the vessel and the perpetual shower of dust and coke which in certain states of the wind fall from the funnels, covering the decks with grit, and immediately blinding any unfortunate inquirer who dares to look up or look ahead.

The coaling of the vessel occupied six hours, the

coals being brought alongside on lighters and tipped down the bunkers amid clouds of dust, which, borne away on the hot, dry wind, literally covered the whole ship, and every one who remained on board.

We only remained one day at Saint Vincent, but we were now at last really in the tropics, and all was so new and full of interest, that even on these barren volcanic rocks it appeared to me that there was more than sufficient scope to furnish agreeable occupation for a year.

The group of the Cape de Verde lies about 300 miles from Cape Verde on the African coast, in latitude 14° to 17° north. It consists of ten islands, which seem to have boiled up from the bottom of a deep ocean. St. Vincent itself is a bare volcanic rock; walls of lava and hardened ashes rise in a precipitous mountain all round, giving an almost stratified appearance to the ash-coloured, parched, barren contour of the bay. It looks like a volcano of yesterday. Among the few houses of which the Portuguese settlement consists, we came for the first time with great delight upon some palms and some cocoa-nut trees, with glimpses of bignonias and a few other flowering shrubs, mostly planted in tubs.

The large island of San Antao, which lies to the north of St. Vincent, intercepts the north-east trade winds, and forcing them into greater elevation, their moisture is condensed into continuous rains by the cold due to expansion, and a constant cloud rests on the summit of the lofty mountains. This island consequently produces abundance of oranges and other tropical fruits, but having robbed the air of all humidity, its little sister, Saint Vincent, is exposed to constant drought and sterility, and its inhabitants depend on importation, even for the most common necessaries of life. They are principally blacks, employed on the various coaling wharves, which are established here in consequence of the convenient position of the island and the excellence of its large and sheltered harbour.

The brilliancy of the still waters in the bay is very remarkable; the coral bottom is visible at great depths, and black divers, in their normal dress, which is merely a wrapper round their loins, come alongside in their canoes, and fetch up every penny or sixpence that is thrown into the sea long before it reaches the bottom.

On leaving St. Vincent other objects of deep interest crowded themselves so rapidly on our notice that a proper record became impossible, and a hurried notice was all that was practicable. Flying fish with their silvery wings now scudded over the water and passed the ship in shoals, taking long flights of two or three hundred yards from wave to wave, often only escaping one foe in the water to meet another in the air; we caught some at night by means of a bright light on the lower deck placed opposite the gangway, which was left open for the purpose; the fish, attracted by the light, flew on board, and were captured and cooked, and an excellent dish they made. They are of the same family as the voracious pike, and use their long pectoral fins more as parachutes than wings.

The Portuguese man-of-war, erroneously called the nautilus, continually flocked by, rising and falling with the wave like a miniature ship. This singular organism the physalia belongs to the hydra family, and is in every respect a jelly fish, with its crested swimming bladder of unusual size floating on the surface, and the long stinging tentacles with which it captures its prey hanging deep in the water below.

The sea was occasionally covered with patches of a red colour resembling a film of red sawdust floating on the water,—probably the microscopic algæ observed by Darwin. Sometimes in the evening or at night we seemed to have entered a sea of liquid phosphorus; streams of brilliant light with bright scintillations were thrown deep under the surface by the action of the screw, and lit up the track of the vessel for a long distance behind. These rhizopods form a large percentage of the whole bulk of the water. I have taken a tumbler full of water from the Menia Straits, near the shore, in which a bottom layer of noctiluca miliaris occupied half the solid contents of the tumbler. At first sight they appeared, like ice, to be only water in another form, as though congealed by vital action, but on closer examination the noctiluca is no less remarkable for its complete and wonderful organization than for its beauty and evanescence. The nights became oppressively hot, with the sky still more or less cloudy, and on the 22nd the temperature of the sea was 81° and of the cabins 84°, with all the ports open. As we had reached altitude 11° 56', and the sun's declination was also 11° 56', we now for the first time experienced the effects of a perfectly vertical sun at noon, so that, like Schemyl, we walked about without shadows. In spite of iced gin-slings and sherry cobblers, and a deshabille almost of native simplicity, it was impossible to avoid constant perspiration, making it difficult to sleep or to write without blotting the paper. Our north-east current had almost died away, the atmosphere was loaded with moisture, and there was no evaporation, and everything foretold our near approach to the Doldrums, which we entered on the 23rd of April, when our trade-wind died away outright,—not, however, without a struggle, for it so lowered the temperature of this region of heated vapour that it expired in a torrent of rain, and filled the sky with dense, heavy clouds.

There was no lightning, although we were in the cauldron in which these north and south currents blend, and partially condense, but when we remember that they are both surface currents over the same waters, and therefore electrically equalized, and moreover that the saturated air is itself so good a conductor as to carry away quietly any little disturbance of equilibrium that may take place, the absence of lightning is satisfactorily accounted for.

The Doldrums follow the sun in its change of declination north and south, but never reach the equator: they oscillate between latitude 2° and latitude 11° north. Hence the north-east tradewinds never reach as far as the equator, while the south-east trades frequently overlap it, so that we met with them on the 24th in latitude 4° north. This is due to the preponderance of land north of the equator, by which the great ascending equatorial current is produced.

## CHAPTER VI.

### ASTRONOMICAL EVENINGS.

THE gorgeous tropical sunsets, with their startling brevity, called our attention to a subject which throughout the journey afforded us the highest enjoyment. I allude to the constantly-changing aspect of the heavens. It was our habit after dinner to pass these lovely warm evenings on the quarter-deck, where we remained late into the night, watching new constellations as they came up from the south, and taking leave of old friends as they night by night sank lower and lower, and disappeared beneath our northern horizon; the heavens really appeared to be passing away, and new heavens were taking their place. Jupiter and Venus, now brilliant objects, climbed so high, rose and fell so vertically into the sea, that they seemed like new planets. The pole star, already dimmed by its lessened height, was sinking out of view, while the southern

## The Celestial Globe.    49

cross, attended by a stately collection of new constellations, attracted our admiration in the southern heavens.

The changes occasioned by our rapid approach to the equator, combined with considerable change of longitude, required some unravelling before they were clearly realized. The rising and setting of the planets given in the almanac furnishing no clue, we had no alternative but to construct a new almanac for every change of position. For this purpose a celestial globe which I fortunately had on board was an invaluable companion. Our globe was rectified daily as our latitude changed, by lowering the northern pole about 3° per day, and the positions of the sun, moon, and planets were daily laid down by means of small wafers correctly placed in accordance with their right ascension and declination. In this manner the position of the planets, as well as the diurnal motion of the whole heavens, were clearly represented.

A long sea journey affords an opportunity for acquiring a little knowledge of practical astronomy which ought not to be lost; a little method will add materially to the enjoyment of the most unscientific star-gazer. The great lines of reference in the starlit sky are the meridian, the equator, and the ecliptic, and their approximate position is so easily determined that there is no excuse for ignorance on the subject.

The pole star is always recognised: its altitude is always the latitude of the place, and by simply turning our back on it and looking due south on our meridian, the altitude of the equator on the meridian will be the co-latitude. A great arc from east to west passing through this point is the celestial equator, which when we are in Lat. 0° passes vertically over our heads.

The ecliptic is the high road of the sun, moon, and planets among the fixed stars, and cuts the equator obliquely. Whenever the moon is visible its position is indicated by the fact that if a line — A B — be supposed drawn straight through the tips of the horns, the ecliptic — E C — will be a great circle at right angles to that line. If any planets are visible they will be situate on this circle.

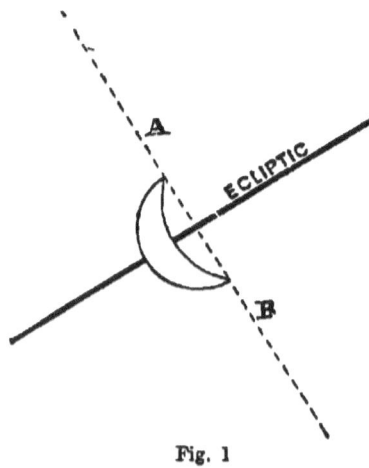

Fig. 1

Although the rising and setting of the sun and planets and other phenomena depending on latitude

## The Almanac.

as given in an almanac will be useless when we change our latitude, the southing of the moon and planets will take place, wherever we are, at the times given in the almanac, subject only to a small allowance for their proper motion in the heavens, due to the difference of time or longitude. For a close approximation in longitude west, add ten seconds per hour of longitude for the stars, and two minutes per hour for the moon, to the time given in a Greenwich almanac; so that if the moon souths in London at 8 p.m. it will south at Buenos Ayres four hours west at eight minutes past 8.*

The moon's age and phases, the times of eclipses, of Jupiter's satellites, occultations, and other similar phenomena, are absolute times independent of our position; they are recorded in Greenwich time, and we have merely to ascertain our corresponding local time.

The eclipses of Jupiter's satellites and occultations of stars by the moon are of great interest to a traveller. It requires a good telescope to see these eclipses, and it is not often possible at sea, but it is a most interesting observation and easily made on shore. The large disc of Jupiter throws a conical shadow into space, and it is on entering that

---

\* The rising and setting are easily obtained from the southing by means of a table of semi-diurnal arcs, or from the simple formula, cos time = tan latitude × tan declination.

shadow at a distance from the planet that the immersion or sudden disappearance of the satellite takes place. Its emersion is equally sudden on the other side of the shadow.

The satellites also frequently pass over the disc and are seen on it as a black speck. The satellites themselves cast a shadow which sometimes passes over the disc of the planet, and is distinctly visible as a dark spot, and it is not uncommon to see both the satellite and its shadow clearly projected on the bright disc of the planet at the same instant. At other times the satellites pass directly behind the planet and are said to be occulted. As there are four satellites, which rotate round the planet in short periods, the first in less than two days, and the fourth in twenty days, these phenomena are of constant occurrence, and afford a very ready means of finding the longitude. For since they are visible everywhere at the same instant of time, like a sudden meteor in the sky, we have only to note the exact time of an eclipse in our own local time, and refer to the almanac to ascertain the Greenwich time at the same instant, and the difference will be our actual longitude without any calculation. An eclipse of the moon is a phenomenon of a similar character, being everywhere seen at the same instant; but this is not the case with eclipses of the sun.

## The Celestial Globe.

Lunar distances and occultation furnish other methods of finding longitude on the same general principle, viz., as furnishing phenomena observable anywhere occurring at absolute epochs of time, and of which the precise Greenwich time is either known or can be calculated.

A few minutes' reference to a celestial globe, properly rectified, solves all the ordinary astronomical problems at sea ; and I cannot too strongly recommend a constant reference to this neglected, but most invaluable, companion. I know of no addition to the luxurious equipment of these splendid vessels that would be more appreciated than a well-illumined celestial globe, daily rectified for the latitude of the vessel; with the addition of some daily indication of the actual position of the sun and planets on the ecliptic. The small globe that I fortunately possessed was always in request, and excited constant interest, not more among the passengers than among the officers of the vessel.

### The Navigation of the Ship.

The invaluable and constant standard of reference in a ship is the compass. But the compass is never employed, as is sometimes imagined, as an independent guide. The popular fallacy that the compass points north and south has, on shore, led to the incorrectness of at least 80 per cent. of all our ordinary weather vanes. About three centuries ago it was nearly correct in England; but the variation changes from year to year, and from place to place. In 1877, a perfectly-balanced magnetic bar, freely suspended in England, was so attracted by the earth's magnetism, as to assume a position such that the north end dipped down no less than 67° 39' towards the north, and its horizontal deviation was 19° 5' to the west of true north. By placing weights on the upper or southern end, and bringing the needle horizontal, the dip of 67° 39' is destroyed, and we have the mariner's compass in the form of an horizontal needle, with a deviation of 19° 5' west. As we travel south and west, the dip and the deviation both diminish; and at Buenos Ayres the needle points 10° to the *east* of true north, instead of 19° to the west. In delicate instruments the balancing of the needle requires to be altered on account of the change of dip, and while I was at Buenos Ayres a tradesman, who had received some excellent pocket compasses from London, was on the point of returning them, as they had been thrown on his hands as defective on account of the needles dipping so much as to touch the compass box. This was, of course, as I explained to him, easily corrected by a drop of sealing wax varnish placed on the lighter end of the needles.

But the needle would be unsafe as an absolute guide, the two compasses alongside each other at the helm differing by six or seven degrees, and these differences vary with the position of the ship.

The actual variation of the compass is ascertained by means of the sun at apparent noon, when the exact degree is noted on which the shadow of an upright pin, fixed over the centre of the compass, falls : the error of the compass, including every disturbance, is thus directly determined. It is equally well ascertained at any hour by a calculation of the sun's azimuth. The compass is only assumed to remain constant during the short intervals between these frequent observations.

The speed of the ship is directly ascertained by the rough method of the log, or by estimation from the number of revolutions of the screw, and the course for short periods is thus obtained with considerable exactitude ; but storms and currents interfere with such a reckoning, unless for very short periods. It is therefore indispensable to ascertain daily the ship's position by astronomical observations.

The latitude is obtained at sea with great facility by an observation of the altitude of the sun at noon, the horizon affording a most convenient datum for the purpose.

## Latitude.

The latitude is simply the observed zenith distance of the sun at noon, plus or minus the declination, according as the latitude and declination are alike or different. We were in the habit of daily ascertaining our latitude in this manner with an ordinary pocket sextant, and it is an operation so short, so simple, and so essential to a traveller, that such a practice is recommended.

The next daily operation on board is the determination of the longitude. This is not quite so simple an operation.

Longitude is the east or west position of the ship with respect to Greenwich. It is measured either in hours or degrees, which, on a globe, are figured on the equator, the whole circumference being twenty-four hours, or 360 degrees. As it is apparent noon at any place at the moment the sun is on the meridian, and the earth makes one rotation from west to east in twenty-four hours, the hours of longitude are practically hours of time; that is to say, if it is noon at Greenwich, then at any spot whose longitude is fifteen degrees, or one hour, east of Greenwich, the sun will have passed the meridian of that spot one hour ago, and the time will be one hour after noon.

Thus the difference of longitude between any two places is practically the difference of time. The problem is therefore reduced to finding at

## Longitude. 57

any given instant the difference between the exact time on board and the time at Greenwich. The Greenwich time is given by the chronometer on board. The ship's time is the angular distance of the sun from the meridian, known as the hour angle. There is no possible direct method of measuring this angle at sea; but, as the sun in its daily course changes its altitude as well as its hour angle, if we determiue the exact relation between the change of altitude and the change of hour angle, so that the one shall give us the other, our problem will be solved; since the altitude can at any time be directly taken with a sextant.

This relation is determined by a problem in spherical trigonometry, which is of such constant use to the traveller that the usual formula is given in the note below.*

The latitude is determined from the observations made at noon, as described. The polar distance, etc., are taken from the "Nautical Almanac." By

---

* $(sin \frac{P}{2})^2 = cos\ S.\ sin\ (S - a).\ sec.\ l\ cosec.\ d.$

When $P$ = hour angle required.
  $a$ = corrected altitude of sun.
  $l$ = latitude of the place.
  $d$ = polar distance = $90 \pm dec.$
  $S = \dfrac{a + l + d}{2}.$

the use of convenient tables for the purpose, the calculation is the work of a few minutes.

The observation for time or longitude is made in the morning or afternoon, when the sun's vertical motion is greatest, viz., when it is near due east or west; while the observation for latitude is made at noon. The ship's clock is altered daily at noon to the apparent time thus found, as given by the ship's bells; but in order to obtain mean time for the regulation of a watch, the equation must be daily taken into account as well as the longitude.

The only other routine observation is the determination of the azimuth, which, when the time is known, is also found in a similar way from the sun's observed altitude, or from an observation at the moment of sunrise or sunset.

These simple observations are all that are necessary for návigation across the Atlantic, and the vessel's position and progress thus obtained, are daily recorded in the log.

## CHAPTER VII.

*VAPOUR AND CLOUDS.*

IT is impossible to traverse these tropical seas without amazement at the prodigious forces that are in operation around us, and without our attention being especially drawn to the extraordinary specific qualities of water itself, in all its forms; whether as a liquid mantle, reducing three-fifths of the earth's surface to absolute monotonous smoothness, or whether streaming into the atmosphere above us in columns of vapour and cloud, or piled around the poles in solid rocks and icebergs. Its great office is the distribution and equalization of temperature; and it would appear to be as specially endowed with qualifications for this object, as materials in the animal economy, such as bone and muscle, are for the peculiar functions to which they are so admirably adapted; although in this case it cannot by any possibility be attributed to development. An exceptional evolution of heat accompanies the condensation of water, from the form of vapour. In the act of freezing, the latent heat set free is greater than for any other known substance; while, even in the form of ice, its properties as a radiator of heat are equally unrivalled. In order to fulfil the important *rôle* for which it appears to be so specially endowed, it departs from the almost universal law of nature, by which bodies expand with heat and contract with cold; it ceases to contract after cooling down to a certain limit attaining its maximum density at 39°. On cooling farther, it begins to expand until it reaches the freezing point. This property, which is essential to the existence of life over a large portion of the globe, is perhaps its most marvellous qualification.

The evaporation of water in these tropical seas is prodigious. It is not even largely interfered with by saturation, for the saturated layers rise as soon as they form, and no high degree of tension is attained. The tension force with which the vapour of water rises, is equivalent to a pressure of no less than $\frac{1}{10}$ lb. on every square inch at temperature 32°. At the temperature of the water in the tropics, say at 80, it is half a pound to the square inch, and under a receiver this tension would be attained before

## Properties of Vapour. 61

evaporation ceased, or the space became saturated. The singular feature of this vapour is, that its expansive force or tension is unalterable, except by change of temperature. If without change of temperature we attempt to compress or expand the vapour, either a portion of it passes back into the form of water, or an additional quantity of water is evaporated; but in the presence of water its tension or density is unalterable. On the other hand, if the temperature is lowered, the vapour immediately adapts its tension to the new temperature. It loses its transparency, and a portion is precipitated in the form of cloud or rain.

In assuming the form of vapour, and *vice versa* in its condensation, water changes its volume to an extent almost incredible. Every cubic inch of water, at a temperature of 80, that is licked up from the sea, produces no less than 44,288 cubic inches of vapour of the same temperature, and adds its tension to that of the air; not that this amount of vapour displaces 44,288 cubic inches of air, for if it did we should in a very short space of time all perish from the absence of air altogether, but, marvellous to relate, it displaces no air at all. The presence of air is totally ignored; and with the exception of the retardation occasioned by the resistance of the air to its diffusion, it attains its limits of saturation, and behaves in every other respect in the air precisely as it would in a vacuum. We have thus two independent atmospheres, the one of air, the other of vapour, and the barometer which records the total pressure really records the sum of two distinct and independent pressures, the one the weight of the air, and the other the tension or weight of the aqueous vapour contained in the air; and change of temperature, which merely expands or contracts the one, exercises, as we have seen, a totally different influence on the other.

The latent heat absorbed in evaporation or set free in condensation, is as remarkable as the change of volume; for at our temperature of 80, every cubic inch of water thus raised out of the sea in the form of vapour, flies away with sufficient heat to have raised its temperature by no less than 1,061 degrees, if it had not been evaporated; and wherever it is transported to and condensed again, this amount of heat is set free.

The mechanical force developed in evaporation is easily calculated, for if a cubic inch of water at 80 evaporates with sufficient force to produce 44,288 cubic inches of vapour, exercising a pressure of $\frac{1}{2}$ lb. square inch, it would, if confined in a tube one

## Weight of Vapour. 63

inch square, raise $\frac{1}{2}$ a lb. to a height of 44,288 inches, or 22,144 lbs., or nearly oneton to the height of one inch.

The weight per cubic inch of the vapour of water, is about two-thirds of that of dry air, at the same temperature and pressure. Saturated air is therefore much lighter than dry air; for though the addition of vapour to a column of dry air must doubtless increase the weight of the column as a whole, it must be borne in mind that the elastic force of the column has also been increased. If, therefore, the surrounding pressure remains the same, the column must occupy a larger space, and although it will, as a whole, be increased in weight, its weight per cubic foot will be diminished.

Another property of water itself is of the highest importance in meteorology. The heat consumed in raising its temperature is greater than for any other known substance; the amount of heat required to raise one pound of water from 32 to 212 being sufficient to raise a similar weight of iron from 32 to 1,652, or to a dull red heat. The prodigious volume of heat accumulated in these tropical seas, can thus be more practically appreciated. The following short table embodies these results, and will, it is hoped, now be understood.

## Vapour of Water.

| At tempe-rature. | Elastic Force. | | Cubic inches vapour from 1 cubic inch water. | Latent heat set free in con-densation. | Weight per Cubic Foot. | |
|---|---|---|---|---|---|---|
| | Inches of Mercury. | lbs. per square inch. | | | Dry air. | Saturated Air. |
| | | | | Degrees. | Grains. | |
| 32 | ·199 | ·1 | 182,323 | 1,092 | 563·0 | 561·6 |
| 41 | ·274 | ·13 | 137,488 | 1,086 | 552·6 | 550·8 |
| 50 | ·373 | ·18 | 102,670 | 1,080 | 542·6 | 540·2 |
| 59 | ·505 | ·25 | 77,008 | 1,073 | 533·0 | 529·8 |
| 68 | ·682 | ·33 | 58,224 | 1,067 | 523·7 | 519·4 |
| 77 | ·909 | ·45 | 44,411 | 1,061 | 514·7 | 509·2 |
| 86 | 1·206 | ·59 | 34,041 | 1,055 | 506·1 | 498·9 |

This subject claims especial attention when the phenomena of evaporation force themselves on our notice on so gorgeous a scale. The elasticity of the aqueous vapour here attains its maximum amount, and its ascent is promoted by its own expansive force. It rises into the atmosphere in ponderous columns, whence it flows north and south, absorbing heat and water in so specific a manner, where they are both superabundant, and bearing them in steady streams to equalize the temperature, supply the rivers, and fertilize the valleys from pole to pole.

The impossibility of equilibrium in an atmosphere composed of both vapour and air, is thus beautifully summed up by Daniell:—

"The earth is surrounded by two distinct atmo-spheres, mechanically mixed, surrounding a sphere of unequal temperature, and whose relations to heat are incompatible with each other. The first, the air, is a permanently elastic fluid, expanding in arithmetical progressions by equal increments of heat, and decreasing in density and temperature, as we ascend in well-known and fixed ratios; but the other element of the atmosphere, viz., its moisture, is not a permanently elastic fluid, but is condensed by cold into the form of water, with a large evolution of that latent heat to which its elasticity as vapour is due. Its elasticity, moreover, increases in force in a geometrical progression with equal augmentations of temperature, and its density and temperature decrease as we ascend in a much less rapid ratio than that of the dry air. Although when mixed with air it ultimately permeates uniformly through the whole mass, it meets with great resistance in passing through the air, and requires considerable time before this equilibrium is obtained. The vapour is thus often forced by the movement of the air to ascend in a medium, whose heat decreases much more rapidly than its own natural rate, and it is thus forcibly lowered in temperature, and condensed in the upper regions into layers of cloud or rain, while its latent heat is set free, and communicated to the air."

The thorough appreciation of this masterly summary is well worth a little attention, and has been the principal motive of this chapter.

## THE DEW POINT.

The Dew Point is that temperature to which the atmosphere must be reduced to become saturated, and deposit its moisture. It may be directly determined by artificially cooling the water contained in any metal vessel until dew is visibly deposited on the exterior; the temperature will then be that of the Dew Point. If the air is already nearly saturated, a very slight decrease of temperature will produce deposition of dew, whatever be the temperature of the air. Heated air, as we have seen, when saturated, holds a much larger amount of vapour in solution than colder air. The dryness of the atmosphere is, therefore, not measured by the actual quantity of vapour contained in the atmosphere, but by the proportion which the amount actually existing bears to the quantity which would produce complete saturation at the given temperature. This is termed the relative humidity; and when we read that the humidity is 67, it signifies that it is 67 per cent. of complete saturation.

The Dew Point is generally determined by observing the difference of temperature shown by two thermometers, one of which is covered with cotton, kept constantly wet. The wet bulb thermometer becomes colder by evaporation, on account of the latent heat absorbed by the vapour as it is formed; but the decrease of temperature soon attains a

## The Wet Bulb Thermometer. 67

limit when it ceases, although the evaporation continues. Augmented circulation from wind will only hasten, but in no other way affect this limit at which the heat imparted by the air must precisely balance the heat absorbed in evaporation.

The difference of temperature between the dry and wet bulb, when it has attained this limit, measures the evaporating power of the air, and as this varies with the amount of moisture, it also gives indirectly the humidity or distance from saturation, and also the Dew Point.

The formula employed for finding the Dew Point involves some calculation. In order to avoid this, I append a set of factors by Mr. Glasbier, by which the observed difference between the wet and dry bulb is to be multiplied, in order to determine the difference between the actual temperature and the Dew Point.

If the temperature of the air, as given by the dry bulb is 53, and of the wet bulb 43, so that the difference is 10°, the temperature of the Dew Point will be 53 — (2 × 10) or 33°. The factor 2 is found opposite 53 in the table.

| Temperature of Dry Bulb. | Factor for Dew Point. |
| --- | --- |
| 10 | 8·8 |
| 16 | 8·7 |
| 20 | 8·0 |
| 24 | 7·0 |
| 26 | 6·0 |
| 28 | 5·0 |
| 30 | 4·0 |
| 33 | 3·0 |
| 34 | 2·8 |
| 35 | 2·6 |
| 36 | 2·5 |
| 37 | 2·4 |
| 40 | 2·3 |
| 43 | 2·2 |
| 48 | 2·1 |
| 53 | 2·0 |
| 58 | 1·9 |
| 67 | 1·8 |
| 77 | 1·7 |
| 94 | 1·6 |

## CHAPTER VIII.

### *FROM ST. VINCENT TO THE PLATE.*

WE crossed the equator on the 24th of April. It was a lovely, hot, tropical day; the water, though scarcely rippled on the surface, rolled by in large smooth waves that raised and lowered two sailing vessels which lay becalmed near us like children in a nurse's arms. Captain Thwaites had told me that in his long experience he had seldom crossed the line without a shower of rain; and although there was no previous indication of any change, at 5 p.m., almost at the precise moment of crossing, a puff of wind sprang up, followed by a deluging shower of warm water from the S.S.E. The welcome breeze filled the crowd of white sails carried by our companions, and, bearing before them a ridge of foam, they disappeared in the cloud and mist. The water had now attained its highest temperature— 83°—which interfered very considerably with our

## San Fernando. 69

speed by lowering its condensing power in the engines.

On the following day, the variable baffling light airs settled down into a pleasant breeze from the south-east, which refreshed the hot ship and absorbed the great masses of cloud that lowered on every side, and we welcomed the agreeable news that we were entering the dominions of the south-east trades, in latitude 3° south.

We passed close along the western rocky coast of the island of San Fernando de Noronha. Our approach to these singular rocks was indicated by the presence of flocks of sea birds, which with few exceptions are confined to the neighbourhood of land.

The island is used as a penal settlement by the Brazilian government, and is so jealously guarded that the officers of the *Challenger* were not permitted to examine its interesting natural history, a lack of liberality that one could scarcely have anticipated among so enlightened a people as the Brazilians. It is about 200 miles from the mainland. The remarkable steep needles of rock which rise like spires, consist of phonolite, columnar at the base, ejected by volcanic action; so rapidly does the floor of the sea sink away from these volcanic islands, that at six-and-a-half miles the depth is 525 fathoms, and at twelve miles the depth is 820 fathoms, falling off rapidly to 2,275 fathoms, where the bottom consists of that wide-spread deposit the globigerina ooze. This singular deposit forms the floor over a large portion of the Atlantic, and consists of the shells of clouds of microscopic organisms, "foraminiferœ," that have lived in the waters above, and which in course of ages have accumulated so as to form a deep bed or stratum. The shells of many species are beautiful microscopic objects, and their complete resemblance or analogy to similar organisms of which the mass of our ordinary chalk consists, has led to the probability that another chalk formation of analogous character is in process of deposition over a very large part of the Atlantic. The enormous thickness of our beds of chalk affords some idea of the prodigious range of time occupied in the deposit of this single geological stratum. These shells are very generally mixed with pumice dust, and even pieces of pumice stone, the undoubted products of volcanic action. The island looked green and fertile, and covered with vegetation, but becomes parched up in the dry season, so as completely to alter its aspect.

On the 26th, the S.E. trades increased to a pleasant breeze, and although not of much assistance, we spread a little sail, which steadied the ship. During the frequent showers which came down in deluging quantities, we were compelled to close our cabins, so that they became uncomfortably hot, the temperature being 85° when we arrived at Pernambuco, where we lay at anchor all night outside the singular reef which runs parallel with the low shore and forms the harbour. This extraordinary wall of sandstone, preserved by its thick coat of barnacles and shells, runs parallel with the coast for a distance of many miles, enclosing a long belt of quiet, deep, brilliant water full of animal life, and, at Pernambuco more especially, tenanted by voracious sharks, attracted by the refuse from the town and the numerous vessels in the harbour. If an unfortunate boatman or sailor falls overboard, or ventures upon a bath, he is but too frequently the victim of these voracious feeders. The reef is barely level with the surface of the sea, which breaks over it in columns of foam, presenting the singular phenomenon of a wall of spray and breakers as far as the eye can reach in the open ocean, with a quiet lake inside.

The opening in the reef, through which small vessels pass on entering the harbour, is limited in width, and the great Atlantic waves roll through with terrible violence, and when there is any wind the whole operation of landing from the steamer is difficult and expensive.

The embarkation out of the steamer into a small

boat is effected in bad weather by means of a chair or cradle, slung from a derrick on the ship; the distance to the town is then a mile or more, through a heavy sea, ending with a sudden rush through the reef on some high wave. All this is by no means agreeable when we are aware of the numerous sharks that are hopefully waiting for a disaster; but the result, when an European lands for the first time in a tropical climate, repays all the risk and difficulty.

We took a trip in the tramway over the now old bridge which was designed and erected by myself in 1856, although this was my first personal inspection either of this work or of the lighthouse at Orlando Point, which was designed a little later.

It is hopeless to attempt any description of the luxuriance of a tropical flora, or of the intense wonder and enjoyment with which a naturalist first revels among its gorgeous charms. The first imposing objects were the great cocoa palms, crowded with fruit, and the numerous other splendid specimens of palms, cycads, and flowering trees with which the public squares and private gardens are crowded. Large lizards, and a small, brilliant-plumaged kind of wagtail crept in and out of the plastered walls. The quays are partially shaded by a row of a luxuriant species of ficus, with large dark, polished, laurel-looking foliage, which effectually intercepts the fierce rays of a nearly vertical sun, and shelters a host of black dealers in gorgeous parrots and other tropical birds and animals, and crowds of handsome, half-naked negro women, carrying heavy loads of strange fruit and vegetables.

We left Penambuco at 5 p.m. The cabins were at 85°, and the night intensely hot. It is more especially during the night that the high temperature is so much felt. The sky was full of heavy clouds, which during the night were constantly condensed by some higher, cooler current, and fell in drenching showers. The air was thus saturated with moisture exactly resembling the atmosphere of a pinery or orchid house at Kew. The passengers all more or less felt the effects, and were anxious to get out to sea again, where the trade winds beyond the reach of the disturbance caused by this heated coast, resume their normal flow.

We were not long before we again encountered our south-eastern friend, now and then rising into little puffs, but constantly light, warm, and pleasant, with enough rolling swell to remind us we were at sea.

On the 28th the night was lovely, Jupiter, Venus, and the moon rising vertically out of the sea, while the southern branch of the Milky Way, rich in stars, rose prominently into view. On the 29th, about 7 a.m., we anchored in the beautiful Bay of Bahia, about 13° south of the equator. Having a few hours to spare at Bahia, we were enabled, through the courtesy of Mr. Tiplady, to make a little excursion on the Bahia and San Francisco Railway. The line for a considerable distance runs through a mangrove swamp, with a dark, dense foliage, and its construction under such a sun was attended with considerable sacrifice of life. The country all round is level, but is remarkable for its wild luxuriance, which speedily obliterates all the works of man, unless maintained with incessant labour. This overwhelming vegetation descends right into the city. The line is everywhere closely hemmed in on each side by an entangled jungle dotted with stately trees, amongst which gigantic palms, the bread-fruit tree, the jaca, orange trees, cocoa nuts, bananas, mangoes, and tree ferns are the prominent objects. Some faint idea of the details of this vegetation may be obtained by a walk through our large hothouses, but the marvellous beauty of the whole scene surpasses all description. The lovely flowers, all of new types, the great tufted palms, the glossy fresh laurel-like green of the foliage, and its wanton superabundance, totally bewilders the lover of nature, while gorgeous butterflies and humming-birds flutter from flower to flower. Marmosets and monkeys innumerable play among the branches, and screaming parrots and macaws in rainbow beauty pass in flocks through the air. How few are aware, as Darwin observes, that within a few degrees from our dreary winter fields, the glories of another world are thus opened to us. Well may such a naturalist have stopped again and again, as he says he did, to gaze on such beauties, in the vain endeavour to fix them on his mind for ever.

The noble bay is scooped out of a granite formation, which extends along the coast for 2,000 miles, and enriches the interior with its fertile sands and *débris*. It seems incredible that with such a soil and climate any competition could be possible, but the growth of the sugarcane is nearly abandoned, and the beetroot sugar produced so slowly and laboriously in our cold soils has driven it from the market, on account of the cost of labour under such a burning sun. The general level of the country is about 300 feet above the sea, and the town of Bahia is partly on this high level, but a convenient lift is constantly at work, raising passengers to the upper level for a small toll, thus avoiding the fatiguing ascent through the various hot and steep and dirty streets that lead to the upper town. The quay at Bahia, as at Pernambuco, is encumbered with produce and with dealers in marmosets, macaws, parrots, etc.

Among the numerous pests that mitigate one's

enthusiasm in the tropics, jiggers and mosquitoes play an important part, but the ravages of the white ant baffle all control. They eat up everything; no plant or flower can be grown in the town unless surrounded by a little pool of water kept constantly filled, and for this purpose a little earthenware trough surrounds every plant. But their ravages are not confined to plants; saddlery, books, clothes, portmanteaus, are all consumed alike. A valuable collection of plans and drawings connected with the railway was entirely demolished, and even the carriages themselves are rapidly destroyed, unless in constant use.

We left Bahia on a lovely evening, bewildered with the crowd of new impressions that so short a visit had imparted. Masses of cumuli, with the edges brilliantly illuminated by the new moon, drifted past us, hurried away by the temperate southern breeze; but a slightly rising barometer, with a much stiffer breeze at lower temperature which set in during the night, reminded us that we were running south and leaving our European summer behind us. The sea was rough, and the vessel rolled about rather uncomfortably, with all the ports closed. The passengers were out of sorts, which is commonly the case after a day on shore in such intense heat. The sea fell to 78° and the air to 76°. This sudden decline of temperature is said invariably to occur on passing the Abrolhos rocks. The temperature undoubtedly does not decline regularly on leaving the equator, but *per saltum* at certain spots, as at Lisbon and the Abrolhos, and again 60 miles from Rio, at Cape Frio, where the change is very marked. At Pernambuco both the sea and the air at 8 a.m. were at 84°, while at Cape Frio both fell to 73°. The temperature of the surface of the sea and of the atmosphere follow each other very closely, and at Cape Frio this sudden phase of cold in both is probably due to the deflection of the Gulf Stream, and to the surging up of an under-current of cold water flowing from the southern seas.

To appreciate this, we must bear in mind that the temperature of the Atlantic generally decreases as we descend, at first rapidly, then less and less rapidly, until we meet everywhere with a thick belt of water at a temperature of about 40°, considerably above that of its maximum density, which is 28°; the maximum density of fresh water is 40°.

The depth of this belt averages 500 fathoms, but is much greater in the deep eastern and north-western basins of the North Atlantic. Below this belt the temperature remains nearly uniform, or decreases slowly till in some cases it falls to 32°. The thinness of the superficial stratum of warm water at the equator is very remarkable, leaving an underlying mass of cold water of vast thickness under this upper hot film. That the vertical sun of the tropics should have so slight an effect, is due to the removal of the heated layer by the trade winds and the absorption of heat by evaporation. The constant humidity, moreover, intercepts a very large proportion of the sun's rays. One of the most interesting features with respect to the distribution of temperature, is the steady increase in the volume of warm water, or in the depth of this upper layer as we move northward. Opposite Monte Video a temperature of 45° Fah. is found at a depth of 250 fathoms. The same temperature is met with at 300 fathoms at the equator, but not until we reach 600 fathoms between the Bermudas and Madeira.

According to observations made by the *Challenger*, a deep under-current of cold water sets in northward from the Antarctic Ocean, at a temperature of about 35°, and supplies the waste occasioned by evaporation from the great surface of the Atlantic; this under-current being interrupted at different heights, and therefore different temperatures, by the configuration of the bottom, the layers or strata which pass northwards are regulated by the height of the intercepting ridges. The deeper or colder strata are thus confined to the

## The Gulf Stream. 79

deeper channels, and are arrested by submarine elevations.

The higher temperature of the surface stratum which overlies the belt of uniform temperature, is due to the sun's heat, affected by many disturbing causes, none of them so patent as the effect of the trade winds, which we have had such ample opportunity of studying ever since we left Lisbon. These winds blowing constantly in one direction, drive the heated surface layers before them in a constant current to the westward, striking the coast of South America about Cape St. Bogota. The stream then divides a northern branch—the celebrated Gulf Stream running round the Gulf of Mexico, while the southern branch runs parallel with the coast of Brazil, and gradually becoming wider is lost in the great easterly drift current in the Southern Ocean. The deflection of the Gulf Stream in the neighbourhood of Cape Frio allows the southern cold current to skirt the coast, and hence, probably, the sudden change of temperature experienced.

The Brazilian coast is backed all the way by bold mountain ranges of granite, which become very prominent as we approach near to them between Cape Frio and Rio, where we anchored in the evening on the 3rd of May.

It is difficult to imagine anything finer than the entrance into this magnificent and commodious harbour, flanked by huge mountains of gneiss and granite rising out of the most luxurious vegetation, sometimes towering into lofty needles or peaks, and sometimes resembling huge castles embedded among trees and underwood in the wildest confusion. The soil, which consists wholly of sands from the decomposing granite, is of the most fertile character, being constantly kept moist from the steaming atmosphere above and the numberless streams of the purest water which percolate between the boulders beneath. The effect of a tropical sun on the vegetation, under such favourable conditions, is totally indescribable. It is difficult to penetrate even a few yards away from the beaten track, and it is only by incessant labour that man holds his own, and retains the limited patches he cultivates, or the roads that give him access to them.

The harbour is well defined by a lofty peak or rock, called "the sugar loaf," and on rounding this peak the town opens out, embedded in trees, and backed by ranges of high mountains covered with tropical vegetation, the summit of the Corcovado being 2,300 feet above the sea. We observed a constant cloud on the top of the sugar loaf, arising either from the low temperature of the rock or from the nearly saturated air from the sea being forced up by these hills into colder strata, where the cloud is constantly formed and again immediately absorbed on the other side, as a stiff breeze was blowing at the time, and the sky was clear. During the whole time we lay at anchor, a similar long fringe of massive clouds remained stationary on the Corcovado and along the whole range of hills that skirt the coast.

The effect of land in producing cloud was strikingly manifest in passing the group of islands at Teneriffe on our return. As we approached the group from the south with a north-east head wind, the shelter produced a calm far out to sea, with an eddy from the south-east. The barometer rose one-tenth of an inch in the eddy. As we left them the north-east wind returned. The nearest island, Palma, was six miles from us, and was an active factory of clouds. They streamed away like smoke from a conflagration, separating into detached masses, with a level base, defining the terrestrial stratum, and slowly dissolved 40 or 50 miles to leeward. Similar clouds flowed away from San Antao at St. Vincent.

We landed at Rio and visited the fine and well-maintained Botanical Gardens, with its magnificent avenue of large palms and fine collection of tropical plants. The heat was intense, but the moisture of the atmosphere was still more remarkable and oppressive. The gardens are irrigated by a small stream with means of flushing, and the ground was luxuriantly green with a coarse wiry herbage resembling turf. The screw pines, sugarcanes, and bamboos, the groves of tangled palms and grasses, the enormous trees covered with epiphytes and creepers, the banana groups, and a host of unknown plants and beautiful flowers in groves, bordered with tillandsias, afforded us a scene that can never be forgotten. The roads are lined with palms, and the houses embedded in their drooping feathers. Humming birds and gorgeous butterflies and beetles roam from flower to flower, orchises encumber every tree, and hang in clusters from every verandah, thriving in full sun in this hot-house humidity. The whirr of the chirping frog "hyla," was incessant, and we were startled frequently by the snapping noise made by the large butterfly, papilio ferronia, which is very common.

In the evening we ascended the mountains to a height of 1,200 feet, by a magnificent road thoroughly well constructed, paved, drained, and lighted with gas, that leads to Tejuca, where there is a well-known English establishment, at an altitude of 870 feet above the town. The views of the bay as we ascend are unrivalled, and the tropical vegetation that covers the precipices on all sides, and lines the road, is indescribably magnificent. The hotel is placed in a valley equally beautiful, watered by copious streams, with excellent baths. The rocky walks are covered with ferns and flowers in endless variety, and it is difficult to conceive any greater enjoyment than an evening in this fairy valley, lit up at night with fireflies, and enlivened by the strange and mysterious sounds peculiar to the numberless denizens of these tropical woods.

The night, however, even at this altitude, was very hot, and the dew was most copious. We were, however, thankful we had not stayed in the city below, where even during the day we had suffered from the stench of the narrow, crowded streets and the want of air, as the surrounding hills concentrate the heat and effectually shut out the refreshing sea breeze which sets in at sunset. After this fatiguing little excursion we returned gladly to our airy vessel, which had now become a home, and it was a relief to escape the heat, and the smells, and the mosquitoes, and cockroaches, and sand flies, and other pests, whose remembrance helps to temper our regret at leaving scenes otherwise so enchanting and agreeable.

Unfortunately, we did not wholly escape the scourge which permanently afflicts this thriving and interesting city. The yellow fever never leaves Rio, but is generally at its height in the summer months. It was causing considerable mortality while we were there, and we had not long left when we found that we had a case on board, which threw a gloom over the journey and terminated fatally a few days afterwards in the River Plate. There is no doubt that this terrible scourge may be entirely avoided, and the extensive works both for drainage and water supply now in hand, must go a great way towards this desirable result.

The deaths from fever had very lately averaged 70 per day. They had now much declined; but at the little port of St. Rosas the mortality was rapidly decimating the small community, and was not confined to any class or locality. The bad drainage and consequent stench, the lack of water, the crowded streets, and the dirty habits of the lower orders, together with the humidity and heat, were abundantly sufficient reasons for the plague, even in this cool season of the year. The most striking feature in the tropics is this great humidity, which tempers to a remarkable extent the direct influence of the sun, so that one lights a cigar with a lens with more difficulty under the vertical sun than in England; but the result is a stifling atmosphere of steam, in which vegetable life seems so marvellously to rejoice. Our sensation of heat depends much less on the actual temperature than on the rate at which heat is abstracted from the nerves of sensation; thus marble appears colder than wood, though both are at the same temperature, and the intense cold of 15° below zero, which we experienced at St. Petersburg, gave no such sensation of cold as a foggy day at a temperature of 35° in England. The dry, cold air in the former case is a perfect non-conductor, abstracting no heat, while the moving humid particles of fog at 35° seize on it with painful rapidity. The dry heat at Arapey, where the thermometer constantly remained at 95°, and occasionally reached 104°, was by no means insupportable so long as we lay still, but exercise was impossible. When we remember that the blood retains a constant temperature of 98°, whether we are in air at 105° in the tropics, or at 0° at the poles, it is far more difficult to imagine by what vital process the external temperature of 105° is reduced to 98°, than to understand the production of heat by combustion of food requisite for maintaining the blood at 98° in the atmosphere at 0°.

We came across a variety of tropical fruits, which are liberally supplied on board; but with the exception of the delicious banana, on which a large population to a great extent depends for subsistence, there is no fruit comparable with our European fruits. We came across several species of orange of excellent quality, but often pithy; and, as a whole, not equal to the oranges we took on board at Lisbon. The sweet lime is insipid, the alligator pear is slimy. The custard-apple has no characteristic flavour, though sweet and agreeable, but not comparable in flavour with the gooseberry. The water melons are deficient in sweetness and flavour. The milk of the fresh cocoa-nut is very delicious, but the bread-fruit, the produce of one of the handsomest trees in the tropics, is insipid, and in no respects equal to the potato or the mandioca, which here takes its place; while the wild pine-apple, which cover large districts, will bear no comparison, except as regards size, with our cultivated pines in England.

The latitude of Rio is 22° 53′ 51″ south, longitude 2° 52′ 36″ west. The mean temperature of the month of May in this latitude is about 73°. At the equator it is 83°, and at Buenos Ayres it is 65°. In England the mean temperature of November, which, as regards climate, is the corresponding month, is 55°. But the mean temperature of the twenty-four hours affords but little idea of the difference of climate or even of temperature, unless we separate it into its two elements, the day and night means; for it is evident that the mean temperature will be 72° whether the day is 86° and the night 58°, or whether the day is 76° and the night 68°; as the sum is the same in both cases. The difference of climate, moreover, depends on many other elements besides temperature; such as humidity, winds, rain, cloud, summer and winter extremes, etc.; and no idea of the marvellous climate of the tropics can be conveyed by a mere record of the temperature.

As we quitted the harbour, with a cool, southerly breeze, we observed clouds at a low elevation round the Sugar Loaf Peak. They had accumulated all day in the north with a little lightning. The atmosphere consisted of a lower warm layer of air of the same temperature as the sea, 74°, with a cooler layer from the south flowing quietly over it. The junction of the two layers was well defined by a stratum of clouds rapidly forming and consolidating. That the terrestrial layer maintained an even surface, so that the upper layer floated on it like oil on water, was evident on looking round the horizon, where the perspective showed the strongly-marked stratified bases of the upper clouds resting on straight horizontal lines closer and closer together near the horizon. Towards the south the whole layer dipped into the sea. The line of contact with the sea was evident, and moved towards us; and at length, preceded by a sudden puff of wind and a rise of the barometer of one-tenth of an inch, it passed over the ship, and a gentle shower fell on the decks.

The sketch Fig. 2 explains this simple and frequent type of disturbance, which will be more fully explained hereafter. The heavier cold current over-riding the warm air made itself felt by raising the barometer. The approach was slow and gradual. It did not advance with a vertical face, but overshadowed and covered a very large area of compressed warm air beneath. Moreover, a nearly vertical tropical sun was shining on the top of this great, cold layer of cloud. Upper layers of cloud were forming at a higher elevation, and these upper layers were drifting north, heralding far ahead the

Fig. 2.

coming mass beneath. These were doubtless the heralds which we had observed yesterday accompanied by lightning, and the Sugar Loaf was a useful measure of the low level at which the air must have been saturated, although apparently free from cloud until it impinged against it. The temperature of the sea was the important element in the whole phenomenon, rainfall occurring almost invariably at the actual line of contact of cloud with the earth.

In the evening the current from the south died

out, and the great stratum of cloud resolved itself into magnificent masses, drifting gently by us, occasionally condensing as they passed into heavy showers; but their peculiarity everywhere was their obstinate retention of the stratified form with horizontal bases,—the thin terrestrial stratum maintaining its impenetrability and its sea temperature. In some cases, while the base was thus truly level, the top was carried away by the upper current, like smoke from a conflagration.

A little later we found ourselves in a warmer and stronger current of water, which gave us an extra knot per hour on our southward course. At night the sea was brilliantly luminous at the depth of the screw, but the air was cold; and the next morning we all closed our ports and took to our great coats and flannels. A red sunset was followed by a brilliant sky, illuminated by a full moon. The temperature was 65°, humidity 88°. The little disturbance we had passed through was fully accounted for by the inroad of a column of this atmosphere which still flowed northward.

We came in sight of the low coast of Uruguay in the afternoon of the 8th, and now the change of climate was but too evident. The atmosphere was dull and cold, the enormous stream of mud and silt brought down by the Plate gave a clayey colour to the water for many miles out to sea; and in a dirty

Scotch mist, with drizzling cold rain, we anchored in the shallow estuary off Monte Video.

In many other respects besides the weather our introduction to the River Plate was unpropitious and disheartening. We had two or three cases of fever on board among the crew, and it could no longer be kept a secret that with the boatswain's mate it was a decided case of yellow fever of malignant type, with no hopes of recovery.

Under the most favourable circumstances the proverbial misery of quarantine in the River Plate is, beyond doubt, the most terrible trial that the traveller has to undergo, and its approach had become the subject of earnest discussion. The quarantine in force when we arrived was limited to fifteen days after leaving Rio, and as we hoped to remain in the *Douro* during the loading and unloading of the vessel, until her departure on the 15th, we flattered ourselves that our purgatory would be reduced to eight days; and in order to avoid the terrible Lazaretto in which passengers of every nationality are all huddled together, I endeavoured to engage a little after cabin in a small tug employed in discharging cargo. But all our arrangements were defeated when the health officer came on board, and learning that we had yellow fever on board informed us that our quarantine must be indefinitely prolonged. We had discharged some cargo at Monte Video consisting of coffee from Rio, tin sheets, soap, and general cargo from Europe.

Monte Video is the only real port in the River Plate, and capable, at moderate cost, of being converted into a maritime centre worthy of the commerce of this enormous river system. It is a well built, handsome city, admirably placed, with well paved streets, and luxurious suburbs. The deep water runs up near the town, and could, by dredging, be continued up to the quays and wharfs. Spacious and magnificent docks could easily be constructed, while the trade of the great rivers, as well as of the districts inland, including the Southern Provinces of Brazil, which have no other natural outlet, must ensure a commercial activity and prosperity of the highest order; and it is a subject of regret to every well-wisher of this fertile little state that such exceptional advantages cannot be utilized. The La Plata between Monte Video and Buenos Ayres decreases from fifty-three to thirty miles in width, but the coasts are flat and uninteresting, with numerous banks of mud and silt, through which the vessel forced its way, leaving dense columns of mud in its wake. The distance from Monte Video to the anchorage, which is six or seven miles from the city of Buenos Ayres, is sixty-five miles. On the 11th of May we came to our journey's end, and anchored in the outer roads of Buenos Ayres, in a thick mist, with a storm of thunder, lightning, and rain.

## CHAPTER IX.

*QUARANTINE AND THE RIVER PLATE.*

ALTHOUGH the health officers from Buenos Ayres had visited the ship, we could obtain no definite information as to our quarantine, but we knew it would be prolonged. The officers of the ship professed to know nothing about it, and they can hardly imagine the value of a little advice under such circumstances, as the passengers are of course entirely bewildered and helpless; some, who had been here before, recommended going back to Rio; my friends on shore recommended me to go on to Patagonia,—either of them was eleven or twelve hundred miles away; others advised returning to Monte Video and undergoing quarantine there; in fact, anything rather than the dreaded hulks at the Tigre, for we thought nothing whatever of the yellow fever on board. We could talk to our friends from Buenos Ayres who came in a steam launch, which rolled about in the dirty rough water alongside, but we could not of course shake hands with them; and it was difficult to hear what they said through the rain and fog and wind, but they could give us no information.

The poor fellow with the yellow fever died during the day, and his death was officially reported. In the evening, the body, rolled up in canvas like a mummy, was hauled up with a block through the hatchway, and placed in one of the ship's boats. The boat was then lowered from the davits and lashed astern of the vessel with a long rope, where it rolled about all night in the moonlight, a melancholy spectacle, which threw a gloom over the ship, and prevented the doctor, who was himself very ill, from taking his hand in the invariable game of whist. Permission for the burial was obtained on the morrow, but the body was to be taken fifteen miles down the river. The boat was towed by a small steam launch containing an officer and a comrade, and the simple procession disappeared in the mist. The service was read over him by the officer, and the body was rolled over with three heavy fire bars enclosed in the canvas shroud. The cabin was fumigated, and all the bedding burnt; but his friends may rest assured that there was no lack of sincere sympathy for the unfortunate sailor thus buried at sea. The other patients rapidly recovered, and it was supposed that the immediate cause of the fever was the carelessness with which sailors will indulge in tropical fruit and luxuries whenever they get on shore.

While we lay in the roads I observed a curious phenomenon with the mist, which rather confirms my explanation of the rain cloud at Rio. The upper surface of what I call the terrestrial layer of atmosphere was singularly defined by a vessel lying about 150 yards away, but enveloped in mist, so that the top-masts were perfectly invisible, and cut off by "the straight lines" at so short a distance from the vessel it had the appearance of a wrecked vessel.

In about thirty minutes afterwards, while at breakfast below, a roaring wind was suddenly heard overhead as the vessel heeled over, and every one ran on deck, where we found a south-west squall sweeping over the vessel from a dense mass of clouds resting on the water in the south-west, followed by a deluge of heavy rain. The hulks and launches discharging cargo alongside were cut adrift as rapidly as possible, and made off to avoid the heavy waves. It was, however, very soon all over. This was my first introduction to the "pamperos," or sudden squalls peculiar to this latitude, and was immediately succeeded by an atmosphere of remarkable brilliancy, the distant coasts with all the shipping of the Plate coming vividly into view. The fog had entirely disappeared, and the close, muggy atmosphere was instantly replaced by a breeze of delightful freshness.

As the south-west cold cloud formed by the pampero swept over us through the saturated air, A was the ship with the straight lines cutting off the upper portion, B was the *Douro*.

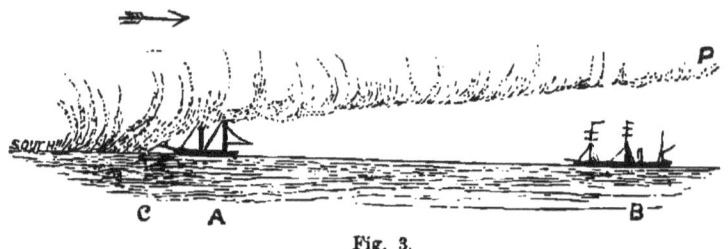

Fig. 3.

The sudden squall arose as the point of contact C swept by us, and the heated wedge P C B was suddenly condensed.

This great condensation creating a partial vacuum and always preceding the wind, must supply mechanical force by which these violent winds are recruited and retain their velocity over such enormous districts, for they could not have left the Andes or the Antarctic with sufficient initial force to travel over the whole continent of South America, and their capricious course indicates that they are not cyclonic.

On the 13th a very small steamer came alongside, and was rapidly filled with merchandise piled up high on the decks, and when she was loaded as deeply as it was possible, we were politely informed that this was the craft that was to convey us also to the dreaded quarantine hulks, where we were condemned to remain until the 27th. Our trunks and luggage were piled up amidst the general cargo on the open deck, for there was no cabin or shelter, the only little space beneath the deck being the small cabin occupied by the skipper and his wife. On getting on board it was difficult to find standing room amidst the cargo. The wind was bitterly cold, but fortunately the rain had ceased; it is otherwise simply terrible to contemplate what must have been the suffering of such a crowd of passengers, including several ladies, exposed in an open sea, without shelter or provisions of any kind, during a journey which lasted until late at night; but it was after arriving at the hulk, perished with cold and hunger, that we felt bitterly our situation. In fact, nothing can exceed the cruel barbarism of our reception and accommodation. No convict in Europe could have worse quarters; the hulk was an old abandoned steamer in the most dilapidated and filthy condition. It had been a paddle wheel troop ship, but the paddle-boxes were broken in and the decks rotten. And when we came alongside on a dark night it was a difficult and dangerous operation to get on board, without ladders or any assistance. There were no lights, no order, no authority, the vessel was in charge of only three common Spanish sailors. There were no beds, towels, nor provisions, and we had at midnight to beg of the sailors to give us a few sea biscuits and a piece of hard native cheese, for which they charged 4s. per head. The vessel had had berths in the cabin below, but they were now in a most filthy condition, and in most of them the bottom planks were rotten or gone. There was not a sheet nor blanket on board, until some wool mattresses and small sheets and blankets were brought about midnight from some other vessel. It was useless to think of sleeping, for the vessel, lying as it did among the islands of the Tigré, simply swarmed with mosquitoes, as well as rats and mice. By the light of some flaring lamps we got out some of our baggage, in the shape of cloaks, and some towels and soap. It was bitterly cold, and we put the ladies in a small cabin by themselves, where they at last sat down, in the greatest distress and suffering, one of them being in a situation in which such exposure was dangerous. We were, moreover, insulted by the Spanish sailors, and this terrible night seemed endless. We all longed for daylight, and seriously feared that our treatment would bring on the very fever that it was intended to guard against. It appeared to us very cruel that we should be thus entrapped without being warned by any one of the fate that awaited us, in which case many of us would certainly have refused to leave the ship and have avoided Buenos Ayres altogether.

First impressions are very lasting, and many whose stay in the country was short will carry away a very damaging and very erroneous opinion as to the civilization of the country, based on the incomprehensible treatment they experienced; but there is no doubt that the government at Buenos Ayres, in this matter of detail as in so many others, had no conception of the suffering we were subjected to, and that they did their best to mitigate our position as soon as they were aware of it. It is, however, difficult to understand why better accommodation was not afforded us between the *Douro* and the hulk, for the scene we afterwards witnessed when a crowd of French passengers was brought on board, who had been all night exposed on an open deck during a storm of wind and rain, was simply heart-rending.

In the morning after this night of horrors, some provisions were sent from a neighbouring quarantine vessel, consisting of soup, beef, and partridges, together with a cook and his assistant.

We all rose terribly disfigured by mosquito bites,

and much out of sorts, but the sunny sky and cheering warmth enabled us to look more calmly on our troubles. It was arranged that our tariff was to be eight shillings per head per day, for two meals,—a very serious charge for many who had barely enough with them for their journey. I, however, sent to Buenos Ayres for little extra luxuries, and after mounting my thermometer and a wind vane, I commenced an investigation of the singular and interesting locality in which the vessel lay.

## THE RIVER PLATE.

The Rio de La Plata, or the River Plate, as it is called, is in no sense a river, but a vast shallow gulf or inlet of the sea, into which two large rivers empty themselves. It is slowly filling up from the enormous bulk of mud, sand, and silt brought down by the Parana and the Uruguay. Its width at the coast is 130 miles, diminishing to twenty-five miles at its upper extremity, which is 180 miles from the coast. The body of water which comes down from these great rivers maintains an average depth of eighteen feet in the Plate, but the depth increases gradually from the shallows among the banks and islands formed at the mouths of the rivers down to the ocean and out to sea. This large area is everywhere intercepted by enor-

## Silting of the Estuary.

mous shifting banks, consisting of silt, sand, and mud. It furnishes a most instructive example of the modern formation of alluvial plains in active process. It was but yesterday, geologically speaking, that this gulf extended many hundred miles farther inland, and so rapid is the process by which it is being filled up by the formation of new islands and new banks, that a very short period will be necessary for its complete obliteration and the projection of the delta of the rivers far out into the Atlantic. The bulk of this deposit is brought down by the Parana, which drains a watershed extending almost to the equator, and supplies a constant and enormous volume of water, highly charged with sediment from the alluvium through which it passes.

The Uruguay, though at times an enormous river, has more the character of a mountain torrent, and is often reduced to a moderate stream, and running generally over a rocky bed its deposit is small and spasmodic.

The present delta of the Parana occupies an area 200 miles in length by twenty-five in mean width, and consists of innumerable large islands of alluvium and silt, undergoing constant change, and slowly advancing farther into the estuary of the Plate.

The process of these formations is very simple; the erosive power of water in any current increases with the depth; thus a constant deposit is always taking place on banks and shallows, while deep channels tend to become deeper. The *régime* is from time to time rudely broken up by floods, when new banks are formed, and large areas of land are raised higher and higher in level. This process is farther increased by materials brought down by the great floods to which the interior of the country is liable. The river thus runs through vast level plains of soft alluvial soil, shifting its position bodily as it meets with harder or softer banks, and thus roams laterally from side to side over enormous valleys. When it comes into contact with the high-lying plains of the pampas, which are eighty feet above it, it forms steep precipices of silt and tosca by scooping away the base and allowing the upper portion to fall in, in vertical sections, when it is speedily carried away by the stream. It thus constantly increases the width of the great alluvial valleys to which it is confined. The banks or precipices thus formed are called "the Barancas." They run for a thousand miles up the river at every distance, sometimes, as at Rosario, in actual process of formation, and sometimes bounding the great plain many miles distant from the present river bed. They are always picturesque landmarks in the river scenery, generally covered with dwarf shrubs, and inhabited by numerous foxes and other animals, and colonies of birds of prey and parrots.

The river Parana, with all its changes, constantly maintains two or more main channels, down which the bulk of its waters flows. The two great channels by which it, at the present moment, enters the Plate are the Palmas or Southern Channel, and the Guaza or Northern Channel, which is the largest and deepest, and forms the great navigable continuation of the Parana into the Plate. It was in the Parana de las Palmas that we were doomed to undergo our quarantine.

It is not directly accessible from the Plate, on account of the shallow banks in course of formation at its mouth, but vessels of small draught reach it by a channel through these banks formed by the River Lujan, which runs into the Plate at the Tigré, and thence by passing through one of the various tortuous channels, or *arroyos*, between the islands. Access to the Parana is now obtained by means of a railway from Buenos Ayres to Campana, a distance of fifty miles, where one of the finest ports on the Continent has been constructed, with wharves over 650 yards in length, and with a depth of sixty feet within a stone's throw of the quay. The width of the channel at Campana is not less than 306 yards, and as the numerous vessels which carry on the river traffic sail by, it is difficult to believe that this is only a temporary secondary channel of this gigantic river. The numberless islands that form the delta of the Parana are thickly planted with trees, consisting principally of the weeping willow and peaches and the ceibo.

### TIDES OF LA PALMAS.

Our position in a floating observatory rendered tidal observations very difficult. The lunar tides are so small and so dwarfed by the effect of wind that they were scarcely observable where we lay. The river rose and fell many feet, according as the shallow waters in the Plate were driven back by south-east winds or swept out to sea by north-westers. Nevertheless there is a regular lunar tide in the Plate, which, though feeble, extends in secondary and even tertiary waves far up the river itself, so that it is high water at one spot and simultaneously low water at another.

Tidal phenomena are brought much more prominently before us in channels, where the configuration of the land obstructs the great ocean tidal wave, and where long and narrowing inlets of the ocean concentrate the tide and prodigiously increase its height. A remarkable instance of this effect, more especially when reinforced by strong wind, occurs in the Bay of Fundy, where the tides rise sixty or seventy feet, and advance with a steep wave or bore. In the open Atlantic, and all along the uniform coast of South America, the tides, like many other natural phenomena, are unaffected by local influence, and are especially normal, and excite but little attention; but they are none the less interesting and instructive.

The tides are produced by the attraction of the moon and sun, and their height varies with the distance and position of these orbs, the effect of the moon being always about three times as great as that of the sun. At the new and full moon the combined effect of the two attractions produce the spring tides, which at a maximum, *i.e.*, at the equinox, raise the waters of the open ocean no less than nine feet. The highest tides, however, are not on the day of the full or new moon, but a day and a half, or thirty-six hours, or three tides later, the tide at any time being due to the position of the sun and moon taken thirty-six hours back. There is thus at all times a meridian two hours, or 30°, east of the moon, when it is always high water. To the west the tide is flowing or rising. To the east it is ebbing or falling, and at right angles it is everywhere low water.

The tides are therefore regulated not by the solar but by the lunar day, whose length is 24 hrs. $50\frac{1}{2}$ min. The tides are thus every day later than on the preceding day by $50\frac{1}{2}$ min., or about three-quarters of an hour. The mean interval between two high waters is thus 12 hrs. 25 min. It varies between 12 hrs. 19 min. 28 sec. at the springs, and 12 hrs. 30 min. 7 sec. at the neaps, and the subsidence or height during the fall is as the square of the times from high water. In charts in which soundings are given, the depths indicated are depths at low water, derived from this relation by observations made at any time.

The time between the passage of the moon over the meridian at any spot and the moment of high water which follows at the period of the equinox, is called the "establishment of the port." In the River Plate the establishment of the port is six hours, or, as a rough rule, it is high water whenever the moon is on the horizon.

The mean time of the moon's southing in any longitude west of Greenwich may be found from the time of its southing at Greenwich by adding 2·1 min. per hour of longitude.

The time of high water will be found by simply adding to the given establishment E the time of the moon's southing, corrected by means of the adjoining short table.

# The Tides.

| Time of Moon Southing. | Moon's Semi-diameter. | |
|---|---|---|
| | 14' 30" | 16' 30" |
| P.M. or A.M. | Min. | Min. |
| 1 | Deduct 17 | Deduct 15 |
| 2 | ,, 31 | ,, 36 |
| 3 | ,, 44 | ,, 55 |
| 4 | ,, 55 | ,, 72 |
| 5 | ,, 60 | ,, 79 |
| 6 | ,, 56 | ,, 72 |
| 7 | ,, 32 | ,, 37 |
| 8 | ,, 1 | Add 9 |
| 9 | Add 14 | ,, 32 |
| 10 | ,, 15 | ,, 34 |
| 11 | ,, 7 | ,, 23 |
| 12 | Deduct 4 | ,, 5 |

If the sum is greater than 24·49 or 12·24, subtract these times; the remainder is the time of high water in the afternoon of the given day.

*Example.*—March 3, 1877. Required the time of high water at Buenos Ayres. Establishment E = 6 hours.

|  | h. m. |  |
|---|---|---|
| March 3, Moon souths at Greenwich | 2 51 | a.m. |
| Add for long. (4 hours) | 8·4 | |
| Moon souths at Buenos Ayres | 2 59 | a.m. |
| Table correction 2·59 hours. Semi-dia. 15' 30" | 49 | |
| | 2 10 | |
| Add Establishment | 6 | |
| High water, March 3, at Buenos Ayres | 8 10 | a.m. |

The altitude of the high tide above the mean

level of the quiet sea is double the depression of the low tide below this level.

Near the mouth of the Las Palmas where we lay we were only a few hundred yards from the nearest islands, which were intercepted by extensive banks of mud covered with rushes, and the opposite coast was about half a mile distant. We had about 18 feet of water beneath us, full of fish. We could scarcely lower a bait without catching the abundant, but useless, cat-fish, a species of barbel with appendages to the mouth, but with dangerous long sharp spines on the fins. One of the passengers, who kicked one of them out of his way as it lay on the deck, received a deep and very painful wound, the spine piercing through a thick boot deep into his foot. Another species was of a golden-yellow colour. Storks and other birds were sufficiently numerous on the islands, which were formerly infested with numerous jaguars, or tigers, and other animals feeding on the capincho, or river hog, and the otters. The only occupants of these islands were the charcoal burners with their small boats, and a few fishermen, with whom we fraternised in spite of quarantine. Moored near us was the hospital ship, to which we should have been removed if the cruel treatment to which we were subjected had made any of us seriously ill. We might then with tolerable certainty have ensured a fever of some kind, if not the yellow fever; but although some of our weaker prisoners were very ill, they resolutely refused to admit it.

After a few days a doctor was sent on board to take charge of the ship, better provisions were supplied, and good order was maintained. He was extremely kind and attentive, and certainly did all in his power to mitigate our miseries. He allowed some of us to make excursions in the ship's boat among the islands, where we could fish or shoot. He lent the ladies mosquito curtains to protect them from these terrible pests, which rendered the nights unbearable. Our incessant beef, beef, beef, which began to be nauseous, was varied with macaroni, and partridge, and yams, or sweet potatoes and omelettes. The coffee was excellent; but our position was immensely improved by receiving from a friend of mine in Buenos Ayres a hamper containing candles and soap, newspapers and towels, brandy, hollands, and cigars, eau-de-Cologne, and fishing tackle. The whole of these articles, especially the candles, towels, and soap, were luxurious beyond price, and added indescribably to our means of comfort and enjoyment, so that we really began to forget our confinement and enjoy the lovely sunny climate and the hot days, although we were now nearly in the middle of our third winter. This anomaly arose from the fact that we passed the winter of 1875 in St. Petersburg, which we left in December with the thermometer at 15° below 0°. We were in England at Christmas, where we remained until April, 1876, and now in the month of May we find ourselves in the middle of a third winter in the River Plate. Although the days were so fine the nights were cold with heavy dew, which, when the nights were still, was deposited copiously on the decks and bulwarks. The stars were often brilliant, and it was our practice to compute a special almanac for every day and watch their rising and setting. The pocket sextant enabled us to obtain our time, and to make a rough survey of the complicated channels and islands that surrounded us. We always lit our cigars with a small lens, and I was struck with the much greater facility with which this was accomplished here than it was under the vertical sun of the tropics, and also with the great difference in the intensity of the sun's heat even here between one day and another, without apparent reason; but this is doubtless due to the presence of invisible vapour in the atmosphere, which is a formidable absorbent of solar radiation.

Vapour, although in an invisible form, is equally efficient in obstructing terrestrial radiation, *i.e.*, in preventing the earth from cooling down during the night, as in protecting it from the effects of the sun's heat during the day and the diathermacy of the air, as the facility with which it admits the passage of rays of heat is an important element in the interesting phenomenon of dew.

We have already seen that if air contains any humidity at all it may always be cooled down to some point at which it will be saturated, and the vapour deposited in the form of dew, and that this point is called the Dew Point.

When at night the air is clear and diathermous, heat radiates rapidly into space from every exposed surface, but certain bodies radiate more freely than others. Long grass radiates very freely. If we call its radiation 1,000, we have the following table by Glashier:—

| | |
|---|---|
| Black wadding, or long grass | 1,000 |
| Short lawn grass | 870 |
| Wood | 773 |
| Iron | 642 |
| Gravel | 288 |
| Hare skin | 1,316 |

The deposit of dew is checked by any obstruction that interferes with radiation, such as a tree, or even a fisherman's net. It is stopped by wind; the cooled particles of air in contact with the earth are carried away as soon as they are formed, and the formation of a cold stratum of air is prevented.

Substances which collect dew in the greatest quantity are good radiators and bad conductors of heat. A polished metal plate preserves its surfaces untarnished, while a plate of glass is drenched with dew. Metals must be reduced throughout to the necessary temperature before dew is deposited.

It was this property of metal which led me to introduce metal caps for the insulators used on telegraph lines.

As instructively observed by Mr. Glashier, the low temperature of long grass as compared with short grass, depends wholly on the heat from the earth being impeded by the length of the conducting blades of the long grass, and not upon any specific properties of grass.

The surface of water, although it is a good radiator, cools very exceptionally and slowly, not only because its specific heat is very great, but because when a particle on the surface is cooled by radiation it becomes heavier and sinks, and the surface will not be permanently cooled until the whole body of water has attained the temperature of 40°, which is the temperature of its greatest density.

It is thus obvious that dew will be copiously deposited on the wooden hand-rail of the bulwarks and not on the iron anchor, which to the touch felt colder, the sensation of cold resulting from the same cause as the absence of the dew, viz., the conductivity of the iron. A walk round the ship in the evening supplied many curious phenomena connected with dew.

Our position in quarantine afforded good opportunity for meteorological observations, and we kept a careful register. We observed on the 15th May a curious phenomenon: a long low dark cloud appeared in the west, advancing rapidly over a clear star-light sky. It passed over us in the form of an immense well-defined arch of singular symmetry, from horizon to horizon, disappearing in the east and leaving the heavens as clear as before. It was succeeded by a second arch, and then by continuous masses of cloud with heavy rain, but only of short duration. The barometer was not affected. These arches were distinct waves, which passed by like smoke resulting from an explosion, and preserved their form without mingling with the atmosphere. They were probably due to some local disturbance, as any movements of the great polar or equatorial currents, which alternately composed our atmosphere, were always accompanied by barometric change. The most violent atmospheric disturbance could be propelled but a short distance, and would be rapidly lost in the general atmosphere, like a cataract falling into a lake, unless it carried with it the elements of its own propagation. The whole mass of the upper atmosphere was therefore moving with these arches, and the effect was due to the intermingling of strata of diffcrent temperatures, originating probably in the ascent of heated columns swept away from their base below, or from the descent of large masses of cold air from above. Avalanches of cold air are frequent near the Andes and among the high Alps. The temperature of both air and water was 67°.

My interest was excited by these arches, as it appeared probable I was witnessing the actual commencement of the disturbance of which they were the heralds.

The intermingling of the first wave was soon obliterated; the second inroad left haze and cloud behind it; the third produced rain. The rain, by its condensation, and the consequent vacuum and changes thus induced, contained all the elements for its own propagation, and for the production of a storm of indefinite magnitude. The increase or decrease of the disturbance depended only on the conditions of the surrounding atmosphere, through which it now commenced to spread and cut its way; its progress being rapid and its force intensified along the heated and saturated valleys of the great river, while it died equally rapidly away in the dry and cold polar currents that flowed over the pampas.

Our little excursions among the islands in a boat

afforded an opportunity of examining the winter botany. Among the most prominent objects was a plant with remarkably bulbous petioles, which floated by singly and in large masses, with fibrous rootlets hanging down in the water. They were masses of the lovely camalote, or pontederia nymphœfolia, so plentiful in South American rivers, carpeting the water wherever it is quiet. It was now out of flower.

The islands being subject to constant floods, the botany is generally of that showy, robust, and coarse character which characterises river botany. The banks are lined with great eryngiums resembling aloes, which grow in large clumps, their dried flower stems being from eight to ten feet high; the wild celery grew abundantly under the willows and poplars, the underwood was covered with clematises and passion flowers, now in fruit. A gigantic and very handsome form of alisma, and also of sagittaria, is very abundant, but the loveliest plant, when in flower, is another pontederia, with its rich large spikes of mauve flowers and water-lily-like leaves. Various species of jussiæa with showy flowers, belonging to the evening primrose tribe, are very common, and might be advantageously cultivated in England. The large pampas grass grows luxuriantly everywhere, and protects the banks.

There were no water mosses, and very few confervae, and very little for the microscope except some desmidiæ.

Many of the islands are planted with peach-trees for the firewood as well as the fruit, but the most showy tree is the ceibo or erythrina crista galli, when covered with its scarlet masses of dense flowers, which resemble lobsters' claws. It is in the spring one of the most striking objects on the river. It grows abundantly on the islands, but it is a gnarled and bare-looking tree in the winter months. The air plants, bromelias and oncidiums, with their pretty spikes of blue and yellow flowers, grow in large masses among the trees, and, tied on hoops, they flower profusely in every courtyard or garden in the city. There are numerous species of the bromelias "pourrettia," some of them deliciously scented, and this hardy epiphyte would be a welcome addition in our greenhouses at home. Some that I have brought home have flowered abundantly, hung in the open air near London. Among the numerous water-plants there is a very plentiful and pretty gentian, with floating leaves, like our villarsia, and showy yellow flowers, the limnanthemum Humboldii.

The terrible wind which every one dreads here is the northern or equatorial current, which, taking into account the earth's motion, probably comes from the north-western tropics. Having thus passed over a large area of heated and humid land, it comes to us as a damp, hot, poisonous wind, of the most disagreeable character, more to be feared than its counterpart, the dry, cold, and irritating east winds that so plague us in Europe. Fortunately it generally precedes a pampero, or southwestern tempest, which clears it all away, the weather here being a continual contest between these two opposing currents. Hence the common Spanish proverb: "Norte duro pampero seguro." So notoriously depressing and annoying is this north wind, that judges were prohibited from giving judgment during its prevalence, while suicides and crimes of violence of every description are supposed to be developed by its baneful influence.

Our first acquaintance with the wind confirmed all we had heard. The dry bulb was at 60°, the wet bulb at 57°, so that the air was nearly saturated. There was a disagreeable haze with light clouds intercepting the sun's heat to a very sensible extent. This was succeeded by a storm of rain at night, the barometer falling two-tenths and the thermometer rising. An essential difference separates this disturbance from those already described.

The rain as before occurred at the point C, Fig. 3, page 96, where the cloud impinged on the earth, but in this case the storm came from the *south*, although the wind was north and travelled northward in the teeth of the wind. The arrow in the figure therefore requires reversal. This feature is usual only with our thunder-storms in England. It will be further alluded to hereafter. I will only now observe that the disturbance was drawn on by the vacuum produced by condensation like the last, but the humid heated stratum was now the upper stratum. Ultimately the rain passed over, and the sky was covered with dew-cloud, which first cleared again in the *south*. It was followed by a change of wind.

During our quarantine it was surprising to observe with what rapidity people accommodate themselves to almost any circumstances. Even ladies, accustomed to all the luxuries of civilised life in England, and whose position was at first very painful for us, began, after five or six days, to bear their fate without murmuring. They formed themselves into a committee to assist those who were most helpless; they manufactured bags of muslin to envelope our heads at night, and keep away the dreadful mosquitoes; they did something towards improving the cleanliness of our quarters, especially after the arrival of our soap and towels, and, though at considerable cost, for I had personally expended £22 on articles sent from Buenos Ayres, our position was greatly ameliorated, and we were patiently and contentedly undergoing our sentences, when on the 20th a small sailing craft came alongside and discharged no less than eighty French and Spanish emigrants, eight of them being first-class passengers. These unfortunate new guests were in the most deplorable condition. They had passed the whole of a bitter cold night on the deck of this crowded little boat, with no other shelter than a sail and that afforded by their piles of baggage. We however did our best to console and assist them, although, apart from the risk attending the mingling with such a crowd of strangers fresh from Rio, our own quarters were now inconveniently crowded, and our privacy invaded.

Our little excursions in the boat among the islands were a most agreeable resource. We had some delicious botanical rambles, and we could always catch as many catfish as we wanted, and cooked them in the tin pan for baling the boat, over a bonfire which we made in the island. They made an excellent lunch, though without salt or butter. On some occasions we took half a sheep with us, which, being secured on a stake and roasted over the wood fire, was an excellent dish. The stake is at first stuck in the ground at some slight distance from the fire, until the meat is warmed. It is then approached nearer and nearer, and ultimately browned over the hot embers. Large joints of beef, cooked in a similar manner, are a common and delicious dish in South America.

At other times these picnics took the form of a teaparty, but they were always most enjoyable, and helped to keep us in health and spirits.

On the 22nd of May, after a cold, foggy night, we had a bright morning. At 8 a.m. the thermometer was at 43°, the wind north; the air was close on saturation, but the temperature of the river was 59°. The whole of the enormous surface of water visible from our ship was covered with little tongues of fog, rising apparently out of the water, and rolling away before a gentle wind until they were dissolved a few feet above the water. There was no fog whatever on the land. The cause was evident. The layer of air in contact with the water was warmed by contact, and its saturation was completed by evaporation from the water. This saturated layer of air at 59° was thus in direct contact with the gently-moving air above it, which was also nearly saturated, but at a temperature of only 43°. So long as they were perfectly still they maintained their transparency, but a constant formation of visible vapour occurred from the gentle movement produced by the wind, the intermingling being beautifully defined by the tongues or wisps of vapour with clear spaces and intervals when the air was motionless. In a short time, as the sun rose higher, the temperature reached 53°, and this beautiful phenomenon ceased.

This little incident assumes importance from the light it throws on the formation of cloud when strata of different temperature and humidity overlie each other. The formation of the vapour was due to the mechanical mingling effect of the gentle breeze, and it is not difficult to understand the enormous condensation that must ensue when enormous volumes of atmosphere of different temperature and close upon saturation are violently and turbulently intermingled.

The day following, we had a return of a very strong and disagreeable north-west wind, and it was accompanied by a very considerable rise of the river; the air was nearly saturated. Cloud again accumulated in the south with straight lines, precisely as before, and we had a gorgeous stratified sunset. The barometer fell two-tenths of an inch. Curly fleeces of cloud passed overhead, exactly like our layers of vapour on the river, and probably arising from precisely the same cause, *i.e.*, the contact of strata differing in temperature.

So anomalous a circumstance as a constant north wind varying in temperature from 43° to 63°, accompanied by a sudden rise of the river, which must have arisen from south-east winds at sea, indicated great disturbance all around us. It reached us on the 24th, in the form of a pampero from the south-west; but the evening sky was lit up in the east by incessant sheet lightning, while masses of heavy cloud, drifting over a brilliant, clear sky, indicated the great contest that had been fought out in some other field, and this was further confirmed by the heavy rolling water that came in from the Plate and rocked the vessel all night, although the air now was motionless.

It is probable that all through this storm we occupied the centre of a cyclone which has travelled east, as this alone can account for the winds from all quarters being so abnormal in temperature and losing entirely their usual characteristics.

While lying at Las Palmas we had many opportunities of witnessing the constant mirage so common in the Plate, arising from the difference of temperature between the atmosphere and the water of the river, and, consequently, of the stratum of air resting on the water. All around our horizon islands and trees appeared to rest on a plate of polished silver, and to be lifted into the air without any base. This effect of refraction was often more apparent over shallows or heated sand-banks, but in all cases only during a still atmosphere, as the wind immediately disperses the refracting layer and mingles it with the general atmosphere.

At length, after a voyage of thirty-one days in the *Douro*, and sixteen days' imprisonment in the *Rosetta*, the glad tidings arrived that our troubles were at an end, and that a little steam-launch would fetch us away on the morning of the 27th, to take us to the Tigré, whence we could proceed by rail to Buenos Ayres.

This was a special favour, for which we were very thankful, as it is usual to land quarantine passengers at the custom-house at Buenos Ayres, and it gave us opportunity of enjoying a delightful trip among the islands down the channel of the Capitan. It was a most enjoyable journey. There were only five of us in the launch, and it was a lovely day. We steamed through an avenue of weeping willows all the way. We visited in route the charming grounds belonging to the ex-president Sarmiento on one of the islands, on which a large notice is posted: " Welcome to the Shade." The palms and cacti, cannas, calladium, and hydrangeas, and a host of other plants, were even now in perfection. The banks were crowded with great alismas and sagittaria and pampas grass, and the gnarled ombus- and willows, and poplars and peach trees, affording the welcome shade. We arrived at the Tigré in about three hours, and, our luggage having been examined, we arrived in the afternoon at Buenos Ayres, and took up our quarters at the Provence Hotel, which was destined for nearly two years to be our head-quarters in South America.

## CHAPTER X.

### *BUENOS AYRES.*

THERE is no doubt that our quarantine adventure had created great though unfounded prejudice against a country in which such administration could be possible. Smarting under recent infliction my own first impressions were biassed, unfavourable, and erroneous. The traveller who arrives at Buenos Ayres after leaving the gorgeous scenery and climate of the tropics, is dismayed at the first aspect of these great level plains and mud banks, without trees or picturesque beauty of any kind. He is equally unprepared on arriving from some great European capital, the result of the labour of centuries of civilization, to appreciate the difficulties already overcome in founding such a city as Buenos Ayres in the course of only a few years. I will record my impressions as they arose.

## The Streets.

The most striking features to a new comer are the badly-paved, narrow streets, which are generally only thirty feet wide, and more especially the narrow and irregular footpaths, so that in many places two persons can scarcely pass. The extraordinary ups and downs as one walks along are very troublesome, each proprietor apparently regulating the level of his footpath, without any regard to that of his neighbour or of the street. These paths are sometimes three or four feet above the road, while in the lower parts of the town the street is sometimes spanned by a bridge, in order that passengers may cross during heavy rains, when these narrow roads become torrents of deep water. There is a peculiar and invariable smell of ammonia, always more prominent after a short absence from the city; and, worst of all, the roads around the suburbs and quays, on account of the absence of gravel or stone of any kind, are either a quagmire of impassable mud and water, or a deep layer of dust, which is raised into clouds by every passing vehicle or puff of wind. It is also a remarkable fact that while the streets are so narrow, the roads are as unaccountably extravagant in width. To pave this enormous width in a country where stone must all be imported is almost impossible, and the most that could be done would be to pave a strip down the middle.

Many of the buildings are extremely fine, but when placed in these narrow alleys it is impossible to see them with any effect. The new banks appeared singularly misplaced. They are crowded with costly decorations, which resemble monumental tombs under the aspect in which one gets a partial glimpse of them by looking straight up into the sky. The bad drainage is a serious evil, although no city in the world possesses such facilities for drainage and water supply. It is singular that it should have been so totally neglected, but the scale on which it is now commenced is quite indefensible.

These are the more striking imperfections which at first sight obtrude themselves on an European visitor anxious to find fault. On the other hand he must not forget, and I have recorded them with that object, that these are the very subjects that are occupying the attention of a highly-intelligent and educated people, who are quite as well aware of the difficulties and imperfections they have inherited as our grumbling friend, and are surmounting them with all the resources at their disposition.

The latitude of the Mercede Church at Buenos Ayres is 34° 36′ 28″ south, and the longitude 3 hrs. 53 min. 25·5 sec. west of Greenwich. The city is built on undulating land, rising gradually as it recedes from the river. The soil is a solid base of tosca covered with alluvium, affording unrivalled foundations for building and facilities for excavations of exceptional character. The old Spanish system of straight narrow streets intercepting each other at right angles has unfortunately prevailed and rendered impossible the picturesque beauty which such a situation must have otherwise ensured; but the public squares are fine and spacious. The older houses are all low, seldom exceeding two stories, and the old Spanish type of a series of rooms surrounding a central open court, or patio, has been singularly adhered to. The houses and streets thus bear a striking resemblance to the houses in Pompeii, except that they are on a larger scale. The patios are often highly decorated and furnished with shrubs and flowers, and contain invariably rain-water wells or *algibes*, with their ornamental marble copings. This form of house with a central court doubtless originated in the turbulent old times, when every man's house was his fortress, as it avoided external windows and left only the doorway to defend. The buildings are generally very substantial, and elegantly furnished and decorated. The flat roofs are paved with brick, and although the fall is very slight and these heavy brick roofs are carried on timber joints, it is surprising how water-tight and durable they prove. They are largely utilized as available space for recreation, and form a most efficient protection against the heat of the sun. The capacious new prison at Palermo, and the new post office, are magnificent public buildings, and the splendid park at Palermo with its avenue of palms is worthy of any capital in the world. These great public works are the more remarkable when we remember that the whole population of the city, though formerly greater, does not now exceed 140,000 people.

The architecture is almost universally Italian; the buildings are generally of stucco and brick, and Italian workmen, who are numerous here, execute this plastering with consummate skill and taste. This work is excessively durable, on account of the lovely climate and freedom from frost, and far greater taste is displayed in the elegant quintas and country houses that surround the city than in our heavy brick buildings around London. On the other hand such light and airy structures would be totally out of place and uninhabitable in England.

The Italians are excellent workmen. They bear the heat of the full mid-day sun with a very slight protection on their heads. I watched with interest some Italian workmen engaged in a house opposite to us. During the hottest weather they worked from daylight till noon without a pause. We saw them frequently take off their shirts and wring out the perspiration, occasionally steeping them in a pail of water. They drink nothing but water, and their wages, though low, are generally economized so that in a short period they accumulate sufficient to return home with enough to start in some business.

We had a large number of Italian labourers on the railway. Their peaceful, quiet habits and incessant industry were exemplary. Their wages were from 3s. 6d. to 4s. 2d. per day. They worked in gangs of about seventy, each gang having its captain, who supplies them with food, which consists of matte and biscuit at daylight, breakfast at twelve, and supper after dark. The breakfast was cooked on the line. It consisted of delicious soup, or puchero, made of beef, potatoes, batatoes, onions, cabbage, and pumpkin, boiled $1\frac{1}{2}$ hours in a large cauldron, and carefully skimmed; the beverage was water. Each man paid 1s. 4d. to 1s. 8d. per day for his food. The men thus save from 2s. 2d. to 2s. 6d. per day, less the cost of clothing, for they live in tents, and a large portion of the amount is invested weekly by these frugal and industrious workmen.

There is no country in which the descendants of the Spanish race have so completely established absolute religious freedom as at Buenos Ayres. The national religion is Roman Catholic, but the power of the Church is strictly confined to ecclesiastical matters. The women are regular attendants at the churches, but the number of men is extremely small. The influence possessed by the Jesuits led to religious riots a few years since, but these dissensions are now forgotten. No religious procession or display is permitted in public.

The bigotry of the population in Cordova is in striking contrast with this unrestricted freedom.

Buenos Ayres being the focus of a large foreign trade, the absence of docks, or harbour, or even of large vessels, which lie several miles away in the roads, is at first sight surprising, but the construction of a port is surrounded with engineering difficulties of the highest order. The water in front of the city is extremely shallow, with a hard level bottom of sand and tosca, and the increase in depth is so extremely gradual that a large vessel cannot approach within many miles, while even lighters of shallow draught are often unable to come within several hundred yards of the shore. The town pier, which runs out a considerable distance, is often left entirely dry during low water with northerly winds. The loading and unloading of sea-going ships is therefore performed by means of lighters between the ship and the shore. The cargo is then transferred to carts on high wheels drawn by horses, which are not unfrequently drowned in bad weather.

In this rude manner this large trade is carried on, employing a very large population and an immense number of horses and carts. The drawbacks are of course very great; the double shifting of the cargo, the damage to goods by water, the waste, loss, and theft, and the great cost, are a serious tax on the trade. It is, however, not easy to devise any better means of dealing with the subject under present circumstances, for carts must be employed in any case to fetch and deliver the goods, and once loaded they now proceed direct to their destination without further delay, whether it is the lighter or the warehouse. It is evident from what we have already said that the possibility of dredging and maintaining a deep channel in an estuary subject to such incessant change must be a problem that requires most careful investigation, and there are no data at present existing that could warrant any thoughtful engineer in expressing a positive opinion on the subject, although all evidence, as far as it goes, is against its feasibility. The experiment now in hand at the Reachuelo, if judiciously carried on, cannot fail to throw considerable light on the subject.

### The Campana Railway.

Our first trip outside the town was on the Campana Railway. It runs for a considerable distance over the great alluvial plain. The first aspect of this great marsh, with the boundless pampas on the higher level above, almost as level as the marsh below, was full of interest. The ditches on either side the line, dug out for its formation, were pools of water overgrown with rushes and floating water plants, and inhabited by large tortoises, fish, and harmless snakes, so numerous that, during the floods, when they seek shelter on the line, hundreds are killed by the passing train. A very handsome species of large toad, brilliantly coloured, which is an object of superstitious dread with the natives, is also common. The biscacho, a large kind of pampas rabbit, has burrowed the ground on each side, where it lies dry, into deep holes, and at the mouth of almost every hole there is generally a pair of small owls, meditating so deeply on the dangerous advances of civilisation as not even to pass into the burrow as the train dashes by.

These owls, with numerous other birds, often sit on the telegraph wires, and are not unfrequently killed by flying accidentally against them. The ptero-ptero, a large handsome plover, very common all over the Continent, with a spike on each wing, utters its disagreeable cry as it sails past. This bird is the pest of the sportsmen, as they continually fly round him, driving away all his game. They seem very affectionate, and can hardly be driven away from a wounded companion, and, on passing near the nest, they become extremely courageous, darting by close to the dog or even the head of the sportsman, endeavouring to inflict a blow as they pass with the spike on their powerful wing. We came constantly across the skeletons and carcases of horses and bullocks killed by the passing trains, and disturbed the great carancho and other scavengers fighting over the feast.

On the river side of the line the marsh is intercepted by enormous rush-grown lakes, full of water-birds, whirling in flocks through the air, and pluming themselves in the sun, or marching on stilts through the shallow pools in search of fish. Among them are flocks of black swans, spoonbills, cranes, storks, wild turkeys, and clouds of wild ducks, teal, widgeons, and snipe, and smaller birds of every description. The wild duck is a most delicious bird, and extremely plentiful in the marshes, and if they could be preserved in tins and brought here they could not fail to command a high price with the epicure. On approaching Campana the line passes close under the cliff, or baranca, which is ninety feet high, and though the river is now a mile away, the caves and gullies in the hard tosca, and other evidences of its origin by undermining, are as patent as though it were a formation of yesterday. It nevertheless must here be of great antiquity, for within a few yards of the line we came upon a tumulus in the marsh, which was opened while I was present, and in which a very large number of skeletons of the aborigines was found, mixed with a great quantity of rude pottery, with flint implements of very ancient date, as flint tools are totally unknown among the oldest tribes of Indians. The flint or chalcedony was probably obtained either far away in the Andes, or may have been brought from Uruguay. It appears to have been a burial-ground, and from the charred stratum and large quantity of pottery, as well as fish bones found near every grave, it is evident the funeral was at the same time a feast. Some bones of deer were also found, but no other bones belonging to the existing fauna of the pampas. Each body was covered with different kinds of earth, sometimes tosca brought from the baranca close by, sometimes sand, or silt, or alluvium from the neighbouring islands and river. These interesting relics have been carefully collected and preserved in the museum at Buenos Ayres, where they will be more thoroughly investigated. The baranca, or cliff, is here covered with underwood, peopled with foxes, owls, large hawks, parrots, eagles, vultures, and other smaller birds and animals. The acacias and dwarf shrubs are smothered with tropeolas, passion-flowers, clematis, and creeping plants, and the glades between are carpeted with brilliant verbenas, and oxalis, and a host of solanaceous plants, with groups of the dried stems of enormous thistles and other compositæ. These thistles form a very prominent item in the botany of the pampas.

The principal autumn plants, as with us, are the compositæ. The closely-packed flowers, with their woolly involucres, effectually defy the sun, but it protracts the ripening of their seeds sometimes far into the winter. The thistle is the characteristic feature of the pampas; vast areas are covered with them, and as they are from five to eight feet high they form an impenetrable barrier even for a horseman. They are too thorny and tough to furnish fodder, except in times of excessive drought, when millions of cattle perish for want of food and water; but they foster the growth of a little grass and trefoil beneath their shade, which serves to keep animals alive till the autumn rains produce sufficient food to fatten them for slaughter or the saladero. The whole country is in many places a bed of two kinds of thistle, the pampas milk thistle and the artichoke thistle, cynara cardunculus, neither of them indigenous.

The line runs on a high bank through this highly picturesque old river bed, with its deep overgrown ditches on either side, and at length debouches on the spacious railway wharves on the banks of the Parana de las Palmas at Campana.

The extensive raised station-ground, backed by the newly-forming town, are flanked on the one hand by a prominent steep spar of the wooded baranca, which runs off inland, and on the other hand by a grove of weeping willows, flooded occasionally when the river is high. The opposite shore consists of the wooded islands of the delta, tenanted with birds, tigers, pumas, and game, while the broad waters of the river are dotted with grebe and other divers, and continually enlivened by the white sails of the numerous craft that bring down fruits and timber, maize, wheat, mandioca, yerbe, canna, sugar, tobacco, wine, and other produce from the fertile provinces of Tucuman and Mendoza, beyond Rosario, from the grand chaco, from the river provinces of Santa Fé, Entre Rios, and Corientes, and from the prolific tropical gardens of Paraguay and Brazil, 1,800 miles distant, on this stupendous river system.

Thus, instead of the dreary monotony I had anticipated, this little trip to Campana was full of the deepest interest and instruction. It is true there are no quiet wooded hills and vales as in England, no vine-covered terraces and castled crags as on the Rhine, no Alpine gorges nor mountain peaks as in Switzerland, but there is an indescribable grandeur and beauty, vastness and freshness in such scenes, peculiarly their own, that is no less imposing and fascinating, and leaves perhaps a deeper and more earnest impression. One cannot help speculating on the future of such boundless fertile territories, in which whole nations could be absorbed, and which, only a few years ago, formed the hunting-grounds and battle-fields of a few wandering tribes of naked Indians fighting for territory! The horse, and the ox, and the sheep have totally revolutionized the great natural features of the country before their slow and peaceful advance. The tiger and the lion, the fox and the deer, and a host of carnivora, as well as their prey, have melted away as effectually as the savage Indian before the patient labour of the colonist. Forests are already modifying the climate. Large towns spring up. The pampas grass and the thistle is replaced by vast areas of waving corn or maize, the wastes are enclosed, and the cattle domesticated and improved both in breed and condition. Thriving homesteads are dotted everywhere throughout the boundless prairie, until the growing prosperity attracts the capitalist from less vigorous fields. Increased production breaks down the bullock cart, and fills the long railway train, and Campana with her busy wharves is destined to utilize a river system of such extent and magnitude that, after years of diligent investigation, its boundaries and limits are still mysterious and unknown.

In countries subject to tropical storms, the greatest difficulty encountered by the engineer is the protection of his works from the sudden and overwhelming floods which are always creating new channels, and occasionally attain dimensions for which he is totally unprepared from any information he can command. The Campana line, running parallel with the river, intercepts completely the natural drainage, and presented many difficulties; numerous and spacious culverts were requisite across every depression. The country is entirely devoid of gravel or stone, and the banks consist wholly of soft alluvium. The marshes were deep and treacherous, and foundations were often very difficult. A storm of unprecedented magnitude occurred in the month of May, 1877, before the banks were thoroughly consolidated, and the damage done was very considerable. A paper descriptive of this storm was written at the request of engineering friends in Buenos Ayres, and is given *in extenso* in the next chapter.

THE NORTHERN RAILWAY.

Although this railway skirts the estuary of the Plate, throughout its whole length it is not subject to any damage from these inland floods, but it only avoids Charybdis to meet with Scylla, for it is exposed to constant danger from the occasional high

## The Northern Railway. 139

tides in the Plate. The water in the Plate, when driven back by constant south-east gales, often overflows its natural boundaries and occasions great destruction. The remains of vessels driven on to high land and there irretrievably wrecked, are not uncommon objects; and the engineers have had to encounter a very heavy expenditure in maintaining this line wherever it comes near to the estuary, as the waters often break on the shore in destructive waves.

This line is especially a pleasure line, and runs through an extremely picturesque district. The steep baranca, dotted with elegant villas, forms its left-hand boundary, while the lake-like estuary on the right, covered with shipping, is always a charming object. It terminates at the outlet of the Lujan River in a little port at the Tigré, among the wooded islands. It runs through a populous district, including the pretty villages of Belgraino and San Isidro, and is lined with quintas, some of them extremely elegant, and well situated in the undulating slopes of the steep baranca which overlooks the river. The islands at the Tigré, already described, are a very favourite resort for pleasure seekers, and especially for rowing and boating excursions. The gun-boats belonging to the Government are laid up at the Tigré, but sand-banks and shallows shut out all vessels that are not of light draught, and the channels leading to Las Palmas are tortuous and narrow. It was in these waters one of the Government torpedo boats, the *Fulminante*, was last year blown up and burnt, with sad loss of life, and with no clue to the cause of the accident.

### THE ENSENADA RAILWAY.

The Ensenada Railway also skirts the River Plate to the south of Buenos Ayres, for a distance of thirty-five miles, and ends at Pontelara, where a substantial jetty, on which £30,000 has been expended, and which is 1,000 yards long, affords at all times convenient accommodation for vessels of moderate draught to discharge alongside. After the description already given of the difficulties attending the landing both of goods and passengers at Buenos Ayres, the latter always embarking in small boats, and in bad weather exposed to great discomfort, and even danger, it is difficult to understand why this port is not used by river steamers. It would surely add vastly to the comfort of passengers going to Monte Video, and shorten their journey, if they started from this point. The only drawback is its unsheltered position, while sandbanks interfere with direct approach, but the original intention was to run on to Ensenada, where there is a deep and well-sheltered harbour. Unfortunately it is only approachable through a channel, that would require to be dredged at considerable cost among these treacherous banks. There is an excellent suburban traffic on this line, which runs through the pretty village of Quilmes, near which the sandy river beach, backed by a fringe of luxurious vegetation, and commanding fine views of the river, offers a tempting site for a charming watering-place. The entomologist and the botanist will not fail to reap a rich harvest among the sandy glades and lovely thickets that line this pretty coast, and shelter tufts of amaryllis and lilies and other plants that delight in the porous soil.

It is only a few years ago that no tree existed over the whole extent of the pampas, but large tracts are now planted; as we crossed the large plains of grazing land, studded with droves of horses and oxen, through which the railway runs, we passed districts that in a few years will assume a forest-like appearance. The principal tree is the eucalyptus globosa, introduced from Australia, which grows quickly and yields timber of some value. It is moreover believed to possess antiseptic properties. Other common trees are the acacias, and some firs, with poplars and willows. There is no doubt that the climate will be greatly modified and improved by this introduction; the radiation from a forest, together with the shade, materially favour the formation of cloud, thus tempering and equalizing the climate and increasing the rainfall, and it is stated that a slight change is already evident in the neighbourhood of the city, where planting has become very general.

## THE PAMPAS.

The excursion which more particularly impressed me with the extraordinary extent, character, and wealth of the pampas or plains, which occupy so enormous an area and are so singularly characteristic of this southern continent, was a journey over the Great Southern Railway. The terminus at Dolores is 126 miles distant from the city; a branch just completed runs to Azul, which is 200 miles from Buenos Ayres. We now leave the great River Valley, with its baranca and undulating boundaries, and travel over these immense distances in what to the eye appears a perfectly level plain or ocean of land, bounded by a well-defined horizon. It is not, however, so level as it appears, which is soon evident from the large lake-like sheets of water which occupy slight depressions, and the gentle gradients on the line, which at Azul has attained on elevation of 400 feet above the sea. In fact these boundless plains rise gently and uniformly in every direction from the sea, at first at the gentle rate of about one foot per mile, this rate gradually increasing until large plateaux near the Cordilleras attain an elevation of 2,000 feet above the sea, and ultimately terminate in the high mountain valleys of the Andes. Speaking generally, the character of the soil corresponds with the inclination, as though a continent had been formed by some great flow of waters depositing boulders and rocks near the mountains; then districts of pebbles and water-worn stones; then coarse gravels and coarse sands; then finer sands; and, lastly, the clays and fine silty deposits which cover the great alluvial plains that form so large a portion of the whole area of the pampas. When we see the enormous addition of similar deposits still in active course of formation over the vast area of the valley and delta of the Parana, we are irresistibly led to the belief that the whole must be due to some continuity of similar action.

The character of the sands and clays and their composition all lead to the conviction that the whole of these plains are the *débris* of the crystalline rocks of the mighty range of the Andes levelled and sorted, distributed partly by great rivers and partly by marine action, during periods in which the whole has been probably several times submerged beneath the ocean; but it is singular that the submergence and elevation should be so uniform over this large area, in close proximity to this active line of volcanic disturbance. The sierras or hills visible from Azul rise through these uniform deposits, like islands out of an ocean of water. The underlying formation consists of rocks of all ages. The lias at Mendoza and San Juan contains ammonites, terrebratula, and other characteristic fossils, while in the museum at Buenos Ayres there is a fine collection of the mastodon, megatherium, glyptodon, etc., found in the pampean clays. The tertiary rocks form picturesque cliffs on the river banks at Parana, consisting of limestone, with conglomerate masses of the great oyster, " ostrea, patagonica," pecten veneris, and other shells; occasional beds of marine shells are met with on the pampas, as well as sulphate of lime or gypsum, and at some of the stations where mortar made of sand containing this material was inadvertently used, the buildings are literally disintegrated and shattered by the expansion peculiar to gypsum as it hardens. I was some years since called upon to report upon similar serious damage produced by the use of gypsum in the erection of the palace of the Baron James Rothschild, at Ferrière, near Paris. One of the most constant characteristics of the pampas is the existence of underlying thick beds of tosca or clays, hardened by the presence of lime. It forms bastard limestone of considerable toughness, with streaks of vegetable carbon, and appears to be universally present. It forms the nearly vertical cliffs of the baranca, and constitutes the hard bottom of the river in front of Buenos Ayres. It is the foundation always sought after by the engineer, and underlies the alluvium throughout the whole valley of the Plate. It is burnt for building purposes, but makes only a lime of inferior quality.

The surface of these singular deposits over which the line passes consists almost exclusively of grazing land, normally without a tree or a pebble. The farm enclosures, or estancias, are the only landmarks, and are generally surrounded by a few poplars, or willows, or eucalyptus, with an occasional gnarled ficus or ombu. The horizon all round is almost sufficiently defined for astronomical purposes. It appeared very remarkable that a soil in which timber grows so luxuriantly when planted should from time immemorial be so totally destitute of forests, or of any vegetation except a few grasses and a very limited variety of flowering plants; and it led at once to the investigation of the peculiarities of this limited flora. It was evident that in an uninhabited and unenclosed plain, swarming with rodents and vegetable feeders, a tree of any kind must always have been as unable to hold its own as at the present moment, crowded as it is with sheep, oxen, and horses. These plains, moreover, on account of the absence of forests, are subject on an average every ten years to terrible droughts, which destroy the hardiest plants, except those whose seeds are endowed with unusual vitality, or are indefinite in number, such as the grasses, the thistles, and the clovers. During these terrible droughts the animals perish in thousands. In the last great secco the carćases of 500,000 head of oxen and millions of sheep strewed the plains, but before they starved these wretched victims consumed not only the grasses and every vestige of vegetation, but the very roots were torn out of the soft soil with their feet, and if the enclosures which shelter the scanty plantations around the present estancias were removed, every shrub and tree would rapidly disappear. The principal grasses which originally covered the pampas were two species of the tough pampas grass, so silicious and wiry as to be unconsumable by smaller rodents. In all the unoccupied wastes these still cover the soil, but as the land is occupied the horses and oxen soon remove this fibrous and indestructible forage, and the plain is covered spontaneously in a few seasons with rich clovers and pasturage, which continually improve by grazing; but during the dry summers this pasture disappears, and a gloomy dusty desert temporarily takes its place. Trees can only be grown by enclosing the land, and if this were done to a large extent the climate would no doubt be materially changed and the fertility greatly increased. These dry plains during the summer months maintain a high temperature, preventing all formation of cloud or dew, and the saturated equatorial currents pass over without condensation, the spasmodic rainfall being due solely to the blending of aerial currents. The high temperature of the pampas, without humidity, produces the incessant mirage that so constantly obliterates the true horizon.

The short-lived botany of the pampas is of course adapted to its peculiar climate and circumstances; no perennial, like the daisy, or buttercup, or plantain in our pasturage, has a chance of escape from the cattle, the drought, and the heat. The dominant flower is a little yellow oxalis, resembling our buttercup at a distance, producing colonies of viviparous pseudo bulbs of extraordinary vitality, too small and too numerous to become the prey of cattle, and too enduring to be destroyed by drought or floods. The other perennials which hold their own are either poisonous, such as the hemlock, or thorny like the numerous dwarf acacias, or tough and indestructible like the pampas and a few other wiry grasses.

One rides nevertheless over these terrestrial oceans with a freedom and freshness that is delicious. Every hillock and burrow teems with curious animals, insects, and birds; snakes, frogs, toads, and tortoises surround the marshy spots, while troops of spoonbills, flamingoes, and waterfowls wade about the waters. The grey fox, after a cautious and inquisitive inspection, darts into his burrow. The skunk scents the air so that we smell him in the passing train; the armadillo and iguanodon run among the clumps of dried thistles and grasses; the biscacho provides them all with homes by his indefatigable labours in excavation, while the hawks and eagles, parrots, owls, pigeons, plovers, and partridges, the cardinal, and numerous small birds enliven the boundless landscape. Larger game is easily found in the "montes," or thickets, that line the arroyos, where the jaguar and puma prey on their weaker brethren, and often on the cattle and sheep.

## The Southern Railway.

At the time that I went over the southern line the camp was in full luxuriance; the country was covered with herds of cattle, oxen, sheep, and horses in dense groups. The railway, like all lines in South America, is not fenced in, and the destruction of cattle is very great. A silly flock of sheep, or a troop of horses or bullocks, will sometimes, frightened by the engine, take it into their heads to follow one another in a stampede straight across the line in front of the train, in spite of the screams of the whistle. When this happens the result is occasionally very disastrous, and the numerous skeletons that line the railway is an evidence of the frequency of such accidents. The cow-catcher, a strong frame of timber in front of the engine, throws them right and left like furrows in a ploughed field, and effectually protects the train. As fences, if adopted, would prevent the escape of cattle caught between them on the line, it is probable that they would rather increase than diminish the number of these accidents. Cattle are, however, becoming accustomed to the danger, and accidents are more frequent with the special than with the regular trains. The large and commodious stations, crowded with merchandize, and surrounded with bullock carts, and the long and heavy trains, bear testimony to the enormous quantity of produce brought down from these fertile pastures, the receipts sometimes reaching £50 per mile per week. The effect of such a line on the prosperity of the district is evidenced by the large and numerous towns that have sprung up all along its course. Railways must inevitably undergo enormous development in a country where roads are impossible, and where the construction of a railway involves no earthworks beyond the throwing up of a bank of loose black mould, taken from the side ditches, on which the sleepers are laid without ballast. It seems impossible to run a line in any direction through these productive districts that would not speedily be remunerative. This has indeed been fully exemplified in the prosperity of every railway of this character in the province. The lines are all single, with a 5' 6" gauge. The towns, as a rule, are slovenly and straggling, with wide roads or streets, which in bad weather are quagmires of mud. The bulk of the houses, and even the churches, are built with rough brick-work, prepared for plastering, but the plastering is indefinitely postponed, and the appearance is very deplorable. I was much amused, on asking a young lady who had just returned from a visit to England, what she thought of London, to find that she was much surprised that in so wealthy a city so few of the houses should be plastered!

On arriving at Azul the difference of climate was very marked. The day temperature was 68°, the night 32°, and the air dry and crisp. The last part of the journey was performed in a carriage, and the rough drive across the hard but uneven surface of the pampas was very exhilarating. Azul being close on the new frontier line of the Indian territory, was occupied by a military detachment under Colonel Donovan, and his account of the raids and warfare of their savage neighbours was full of interest. We saw a few Indians, apparently of the lowest type of humanity. I was surprised to find so far away from all civilization a powerful flour mill, belonging to a most intelligent Frenchman, driven by a 15-horse power water turbine, supplemented by a 20-horse power steam engine. The machinery came from Paris, and the engine, mounted on wheels, had been brought by horses across the pampas. The fuel used for the engine was wholly pure sheep dung, collected in cakes from the enclosures where these numberless flocks are folded. It is there deposited in deep beds, and afforded an excellent fuel, and at small cost; but it seems a sad alternative, to consume so valuable a manure in this fashion; and fertile and inexhaustible as these virgin districts now appear, the time must surely come when such waste will be impossible. The mill-building and dam were constructed with great skill and solidity, and gave evidence of well-deserved prosperity. The town, moreover, had a thriving aspect, and a branch of the Imperial Bank was doing a large banking business. This hitherto inaccessible district cannot fail to rise quickly into great importance under the fostering influence of the railway communication just completed. We slept in a large and comfortable establishment, with its invariable billiard-rooms. The cuisine is not enticing; the ordinary dinner of five courses of beef always leaves the difficult problem of deciding which was the toughest and which contained most onions, but the appetite, inspired by the clear and brilliant atmosphere of the pampas, furnishes a sauce that any epicure would envy.

The following day we returned to the junction at Altamirano and thence we ran on to Dolores, a total distance of 218 miles; we arrived there about half-past one in the morning. It was too late to find quarters in the town, we therefore ran the carriage under an engine shed, and, rolling ourselves up in ponchos and cloaks, improvised the best bed we could on the carriage seats. The night was cold, with fog, and a little hoar-frost, on the grass, but as soon as it was daylight my more vigorous companions enjoyed a shower or rather a storm bath in the open air, obtaining the water from the tank used for supplying the engines by means of the large leather hoze. I was content with a bucket of water, under more sheltered conditions. We were not in the humour to grumble at the breakfast which followed at Dolores, and we visited the fine new prison, but the roads were so detestable that it was difficult to keep one's seat in the carriage, and I felt certain we should be upset into some quagmire. The fences round the town are formed of the great cactus or prickly pear, which makes a totally impassable barrier. There was the usual proportion of unfinished, unplastered, and unroofed houses and churches and barbarous walls, interspersed with several good houses; the population is about 5,000 people, spread over a large area. The habits of the people in the town are peculiar. Business is done early, from seven to eleven; then breakfast, billiards, and the siesta; another short phase of business from four to five; then the dinner, with billiards and cards till midnight. Drinking is becoming a serious evil. The topic of conversation is generally politics, of the most local and personal type. In the camp, on the contrary, all is downright hard work, with substantial but very monotonous fare, consisting almost exclusively of meat,—mutton or beef. Although so easily grown, the cultivation of vegetables is totally neglected, and it seems incredible that, in a country covered with cattle, butter is almost unknown, the only dairy produce being a very dry, hard, and tasteless kind of cheese. Living wholly on horseback, the settler enjoys intensely the keen vitality and unrestrained freedom imparted by rambling over these boundless pastures under a brilliant sky, and though at first the total absence of what we call comfort is rather trying, very few who have once been accustomed to camp life can be again induced to submit to the restraints and obligations inseparable from higher civilization. I have often indeed been struck with the readiness with which the most intelligent and refined fall back into this apparently natural life. Our high civilization is only attained and maintained by incessant labour and education, but it is like a leaky cistern, kept full by constant pumping, which rapidly empties itself when the labour ceases. In the same manner, the natural instincts resume their sway with surprising readiness when free from restraint, and the habits and civilization which have been the result of years of laborious training and example melt rapidly away in a few short months under the influence of this free and levelling life. Large fortunes are rapidly accumulated by estancia holders, but their wealth has little influence on their habits and modes of life; their income is generally invested in increasing their holdings, and in a country where the finest lands can be bought at 20s. per acre, the extent of land held by the large proprietors is measured by square leagues. With so spare a population, there is of course occupation for every one who chooses to work, and the poverty and distress so common among us is totally impossible and unknown. There are few features of engineering interest in these railways, but there is a fine iron bridge 240 metres long over the Salado, one of the very few rivers that drain the pampas. The absence of rivers over this vast area, with a large rainfall, is a very remarkable characteristic. The rain drains into a few large lakes and arroyos, and excavates large temporary channels in the alluvium, but the bulk of it is absorbed by the soil.

The low-lying lands are thus subject to terrible floods, but it is a peculiarity not easily explained that the waters in the valley of the Salado continue to rise for a long period after the cessation of the rain. I have elsewhere endeavoured to explain this phenomenon in connection with the great floods of May, 1877, during which millions of oxen and sheep were destroyed in this same valley.

We returned from Dolores to Buenos Ayres on the third day, after a most enjoyable excursion, during which we had travelled 544 miles. The distance of 200 miles from Buenos Ayres to Azul was performed in ten hours.

### The Western Railway.

The next trip was over the Government or Western Railway, from Buenos Ayres to Chivilcoy, one of the most thriving and remunerative lines in the country. The sumptuous breakfast and lunch *en route*, provided by Mr. Cambaceres, under whose control this prosperous and admirably administered system is worked, afforded us a day of most luxurious enjoyment, but was not calculated to enrich my note-book. The line passes through a magnificently fertile country, on which one could not fail to be struck with the great development, as evidenced by the large and thriving establishments on all sides. Extensive areas were well planted, and what only a few years back must have been a treeless plain now resembles an English wooded district. This extensive planting cannot fail to exercise a beneficial effect not only on the climate and produce, but on the habits and progress of the country.

The open camp bears the same aspect everywhere, but the quality of the cattle is here obtaining attention, and the fostering influence of a railway is especially prominent in the progress and transformation due to its earlier construction.

The present terminus, Chivilcoy, is a large but straggling town, with the invariably bad wide roads, and the usual proportion of unfinished, ungainly brick buildings. Mercedes is a fine and thriving large town, with every appearance of movement and progress. Nothing would be more erroneous than the opinion that would be formed of the resources of this great country by a visitor who confined his attention to the great centres of population. All our older cities were built along navigable rivers, as furnishing the only means of communication. But without rivers or roads, if the railways had not come to her assistance, it is difficult to imagine by what means the boundless and apparently inexhaustible plains of this great continent could ever have become utilized or inhabited. The speedy progress that characterizes the limited districts as yet accommodated, the rapid rise of large and thriving towns and establishments, the prominent accumulation everywhere of productive capital, as well as the increase of population, betoken a future of wealth and prosperity which nothing but the grossest political mismanagement can delay.

## THE CRISIS.

During the time of our visit Buenos Ayres was slowly recovering from a financial crisis of unprecedented severity, which had ruined many families in the city, and even diminished the population. It originated in the reckless trading and building and land speculations that had been encouraged by the facility with which money was obtained by means of the public loans. The public expenditure was, moreover, far in excess of either the wants or means of the country, and this demoralizing influence had also rapidly produced its natural result. The mischief was almost entirely confined to the cities where it had originated, and the camp generally was only indirectly affected. The rapidity with which these new countries surmount such difficulties can hardly be understood in Europe; the circumstances are entirely different. In a manufacturing country a severe crisis affects the whole community. If the merchant finds no market for his manufactured goods, the manufacturer ceases to produce, the iron and coal are no longer required nor produced, and not only capital itself, but the miner, and artisan, and the labourer are all involved in forced idleness or unproductive employment. But in a country relying on its cattle for its wealth, commercial distress scarcely affects production, for so long as public order is maintained, and taxation is not totally prohibitive, the cows go on calving, and the sheep lambing, and the cattle fattening, and wealth goes on increasing and accumulating without any change in the small amount of labour employed, and independently of the distress or prosperity of the Government officials, or of the city merchant, or even of the state of the market. Under such conditions, if the terrible temptation of making large issues of inconvertible paper could have been avoided, and moderate retrenchments enforced, the return of prosperity would have been extremely rapid. The youth and vitality of the country is in no way more clearly demonstrated than by the fact that it sustains, without much check to its recovery, this expedient, which would be so fatal to our older community. The resources of the republic, economically administered, are, no doubt, abundant for all legitimate purposes, and they are exceptionally elastic and progressive; and the time cannot be far distant when so highly-intelligent and prosperous a community will modify the present demoralizing and ruinous system by which a total population far less than that of London is supporting fourteen independent governments, with armies, and presidents, and ministers, and paid senators, and congress-men, and hosts of minor State officials, while scarcely a town in the republic has got a common town surveyor to attend to the drainage or regularity or cleanliness of the streets, or to minister to the first wants of the most incipient civilization.

The most serious evil is the laxity with which excellent laws are administered, and the total neglect of all the minor duties of administration. The heads of departments are not seconded in their efforts to reduce these grievances; the numerous hosts of employés are as a rule sadly incompetent, and their habits and experience are inconsistent with their duties, while the authorities themselves are too much occupied in vainly endeavouring to satisfy an insatiate host of political friends, on whose interested support they are so dependent. Matters which with us excite amazement appear here to attract no serious attention. It seems to us incredible that a congress can be so constituted as to deliberately vote year after year an expenditure exceeding even the estimated revenues of the country by 40 or 50 per cent., and that this should be a paid congress is still more inexplicable.

The short tenure and the profitable privileges attached to office encourage contests for political power, which undermine honest independence and excite passions that none but sordid interests could arouse, while the population is too scattered and too much occupied with their own pursuits to control the executive by any well-organized expression of public opinion. During my residence in Buenos Ayres a body of voters quietly going to the poll were driven back by a volley of musketry from the roof of a neighbouring building, and what is still more remarkable, the perpetrators of the outrage maintained a right to retain the arms thus used in a time of profound peace.

These anomalies extend through every detail of administration. The stoppage of the public works amounts to a serious calamity. It is of course certain that if the tender of a contractor like Mr. Wythes had been accepted, the drainage of the city would have long since been completed, and yielding a large revenue. His tender amounted to about £1,300,000, whereas the amount already expended exceeds £2,000,000, and not a single house is drained, while the estimate to finish the work is now from one to two millions. A committee of gentlemen without the remotest experience in such matters, and with other interests to consider, could scarcely have arrived at any other result.

I have only alluded to these matters because they are at this moment claiming public attention, and will doubtless be remedied, for there is no feature of higher promise in this country than its patient endurance of such evils and its peaceful efforts to remedy them.

As a rule, foreigners abstain from all interference in public matters, and as the native families avoid practical pursuits, and almost exclusively adopt the learned professions as leading to political advancement, there is a dearth of practical men of business, which is sadly manifest in all the ordinary routine of administration. The intolerable result may be realized by imagining our own town surveyors, harbour masters, municipal and custom-house officials, heads of departments, and even ministers of state, to consist of young doctors or barristers, not only without knowledge of the special duties of their respective departments, but without business experience or habits of any kind, and often with no other object in view than political promotion.

These incongruities are disappearing, and will be swept away before the rapid advances which the country is making; and with its vast and unbounded resources, its fertile lands, and lovely climate, and the solid attractions that such a country must always continue to offer to the capitalist and the emigrant, nothing can long delay the proud position in store for this republic amongst the great rising nations of the new world.

In concluding these notes I must not omit to bear testimony to the proverbial hospitality and urbanity characteristic of the Spanish race, which is everywhere met with in these republics. We left with great regret a very large circle of excellent friends, to whom we are deeply indebted for all our enjoyment and information; and if some of the remarks that have been made may appear harsh or ungrateful, they must be attributed solely to a sincere desire to hasten the rapid strides with which these young countries are expanding into greatness, and to strengthen the energy with which they are themselves so resolutely overcoming the abuses they have inherited.

## CHAPTER XI.

### CLIMATE AND METEOROLOGY OF BUENOS AYRES.

IN a young country where every one is practically occupied with the sterner duties of life, no careful systematic meteorological records are available, and a traveller must depend largely on his own observation.

Although a careful continuous register was never neglected, my observations were made with a specific object in view, which requires a little preliminary explanation. I was speedily convinced of the slight analogy existing between our weather phenomena here and in Europe; on the other hand I was equally impressed with the fact that no country furnishes so fine a field for investigation. In its meteorology, as in its botany and river system and its natural history, there is a vastness and normal simplicity and freedom from local disturbances which prominently lays bare the great fundamental causes and elements of every phenomena. The forces at work are here on a vast scale overriding minor perturbations. The great bare plains, including the grand chaco, extend over 15° of latitude and 12° of longitude, or occupy an area nearly 1,000 miles square. The lower half is totally devoid of forest or trees, while large areas in the more tropical districts are equally barren. To the east we have the uniform surface of the Atlantic at an average temperature of 67°, with all the important influences of water and humidity in full force on the grandest scale. To the west we have the great chain of the Andes, covered with perpetual snow, condensing every passing cloud and interrupting every communication with the Pacific coast. Our southern boundary is the cold Antarctic Ocean, with its floating icebergs; and lastly, on the north we have the active cauldron itself, a boundless region of tropical forest and vapour, the essential origin of all meteorological disturbances.

The only local interference with this singular uniformity is the great river system, which has excavated a wide valley through the plains and furnishes a broad belt of lakes and marshes and islands, all more or less covered with forests and vegetation, extending from the tropics to the Atlantic.

I availed myself of railway surveys and other sources of information in carefully determining the inclination of the apparently level plains of the pampas. The country rises 300 feet in a distance of 700 miles as we ascend the River Valley, but the pampas rise much more rapidly, attaining a level of about 500 feet in 300 miles, and 2,000 feet at Mendoza, within a radius of 500 miles. These general results are interfered with by occasional hills or ridges of rock, which rise through these levels like islands in a lake, and also by depressions, occupied by salt lakes and saline marshes, as though the sea had not yet completely resigned its territorial rights over its ancient bed; but such exceptions are on too small a scale to affect the general uniformity as regards meteorologic influences.

On these great battle-fields there rages perpetually an uncompromising contest for possession between the cold winds from the south and the heated, moisture-laden currents that flow in from the equator. In these contests all similarity to our own climate is destroyed. The great cyclonic movements which follow one another across the Atlantic, and disturb the temperate zone, are here completely merged in higher laws due to the proximity of the tropics. It is indeed highly probable that these cyclones are offshoots of tropical disturbances, and originate among these heated realms as rivers do from mountain streams, the tornado and the great cyclone bearing often much the same relation to each other as the mountain torrent to the quiet river in its valley bed. The northern hot and moist winds come to us, on account of their initial equatorial momentum, as north-east winds, while the cold southern currents arrive from south-west. The change from one to the other is usually sudden and very manifest. The barometric changes are small, but the barometer invariably rises with the cold south wind and falls with the moist warm north wind, and its height is at all times a very fair measure of the relative heat and moisture existing in the actual atmosphere. Even the quantity of rain-fall that may be anticipated is approximately given by the barometer. The uniformity with which the barometer and thermometer move together is very striking; a thermometer, with an appropriate scale, would indeed furnish a very fair make-shift barometer, except that the fall of the one is equivalent to a rise of the other. The general routine of the great periodical storms is extremely simple, and is as follows :—

Accessions of heated air flow southward, and gradually occupy the whole atmosphere of the pampas, in periodic times of singular regularity (see fig. 5), each wave intruding farther than the last, until the heated and saturated atmosphere reaches some southern limit, often the Cordilleras, where condensation sets in and rain and storms begin. The disturbance, once commenced, retreats with accelerated energy, drawn back by the vacuum produced by its own destructive condensation; the cold southern current follows closely in its wake to fill the void, and, in the form of whirlwind and hurricane, produced by the violent disturbance, it *apparently* drives back the stormy clouds, and restores the normal blue sky and refreshing atmosphere so eagerly awaited. It was to these great normal disturbances that my attention was specially directed, as it is always at the actual junction of these opposing currents that storms attain their highest intensity. The first feature that struck my attention was the apparent resistance to mixture which exists between currents differing greatly in temperature and humidity. It is so great that mechanical force is an essential element in any storm, whether in its commencement or progress, and this force is almost always generated either in close proximity to or in actual contact with the earth's surface, *i.e.*, in what I term the terrestrial layer. Strata of clouds, however widely they differ in composition, often quietly overlie each other in the atmosphere above us, even when travelling in different directions, the line of demarcation being only visible in the form of a few wisps of cloud or thin strata of visible vapour. If this equilibrium is disturbed by force, such as the rising of a heated column through their midst, or contact with mountains or with the earth itself, a mechanical mixture is set up by which the disturbance is further intensified; it dies out again, sometimes in the simple formation of cloud or rain, but under favourable circumstances, accumulating in energy, it attains the violence of a hurricane when opposing winds, even from directly opposite directions, may meet and add fresh fuel to the flame.

The focus of a storm is not at any altitude, but at the point of actual contact between the storm cloud and the earth itself. Apart from the largely increased thermal effect due to actual contact, the varying currents are forcibly intermingled, and rapid condensation is promoted by the solid resistance. The anomalous and rapid decrease in the quantity of rain collected by gauges elevated above the ground, is thus satisfactorily accounted for.

One characteristic of the terrestrial layer is its permanent transparency; fog and cloud in contact with the earth are very infrequent except on mountains. As a rule the earth is warmer than the air, and even saturated air becomes transparent by its proximity. It is when the earth is colder than the air that the vapour is converted into fog, or is violently thrown down into rain. The air is a bad conductor of heat; the higher temperature of the earth and of the layer immediately subject to its influence is often to a great extent maintained by vapour and cloud above, which intercept radiation. Hence the anomalous results of all experiments on the rate of decrease of temperature in relation to elevation.

Even at sea this permanent transparency is maintained, except on the equator, but here the water retains a temperature abnormally low; the solar heat is consumed in evaporation, and not in heating the water. It is moreover arrested by the vapour thus formed, which becomes warmer than the water, hence the constant cloud and mist where it would least be expected.

To return to our subject. We have seen that the general circulation of the atmosphere is due to the ascent of heated columns of vapour and air over the whole belt of the tropics; as regards the ocean, this ascent is due, not solely, as on the continent, to the atmosphere of air as a whole, but to the atmosphere of vapour. It is the vapour that rises from the water, with the great elastic force due to its temperature, and drags the air with it; this heated belt, thus perpetually reinforced with heat and moisture, is continuously expanding laterally, as well as rising and overflowing in the upper regions; the lateral limits north and south tend towards continual increase, until arrested by some opposing force. It is this lateral extension that produces the hot and stifling northern wind that perpetually tends to overflow the pampas. It often progresses unopposed until it meets the Andes; this is the magic wand that breaks the spell: not only is all further progress instantly arrested by condensation of the columns that first arrive, but the work is all undone; the vacuum thus occasioned by condensation and the ascent of columns of liberated heat is instantly filled in by the heavier polar current from the south. Like a train of gunpowder, the disturbance once set in rapidly travels north, amidst storm and tempest, until it is lost again in the great belt of vapour from which it emanated. The great object of this aerial campaign is however perfectly accomplished, as testified by all the rills and streams that fertilize the mountain valleys of the Andes, and the flocks that graze over the boundless pastures of the pampas, that would otherwise be a sandy desert. The immediate cause of the propagation of the storm is the vacuum formed by condensation. To restore the equilibrium, the air rushes into the vacuum from every direction, cold columns of air are drawn down from the upper regions, heated vapours flow in from the north, and the cold polar current from the south. The condensation of the vapour by this violent admixture creates a fresh vacuum, and maintains the storm. If the atmosphere as a whole is stationary, the storm travels north, but if as a whole it is moving south, the storm may remain apparently stationary at the same locality. The vacuum is partly due to the void left by the condensed vapour or the reduced tension occasioned by condensation, but it is principally produced and maintained by the ascent of columns of air raised to high temperature by the prodigious quantity of heat set free. It would appear to me highly probable that this heat is reinforced by electric agency. The potential of the electricity in these masses of condensing vapour is lowered by its escape to the earth as it is set free, and all reduction of potential is invariably accompanied by a corresponding liberation of heat. In the form of lightning this liberation of heat is evident enough, but the flow of electricity through this vapour in an invisible form must also be very large.

Although these are the general features of such storms, there is considerable variety in the detail, more especially as the expanding equatorial atmosphere more frequently overrides the terrestrial stratum, and hence the singular anomaly that, although the wind is north, these storms almost invariably originate in the south, and travel directly against the wind. The following figure (fig. 4) will explain this puzzling feature.

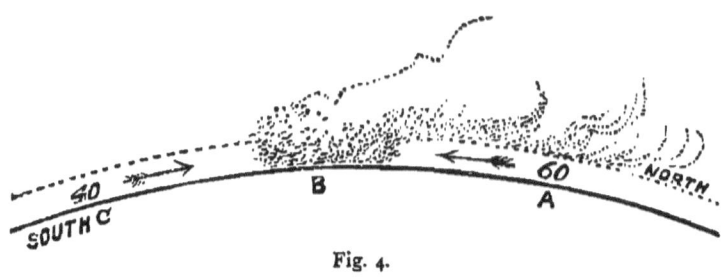

Fig. 4.

A B C represents the earth surface, the north, or the tropics, lying to the right. B is the focus of disturbance when the heated columns from the tropics, now in full retreat by condensation, break up the terrestrial stratum and impinge on the earth. The winds being drawn into the vacuum produced by the storm will be south at C, and north at A; a spectator at A, therefore, sees the storm approaching him from the south, and against the wind, and when it has passed here the wind shifts to the south, and his thermometer falls from 60° to 40°. The sky, moreover, clears up again in the south.

The marshalling of the great aerial forces preparatory to a storm is a grand phenomenon. First, light scuddy clouds like skirmishers ride rapidly past, condensing and reforming as they drift along.

They mark the confines of the terrestrial layer, like tongues of fog upon a lake; they are ill-defined in the zenith in plan, but on the horizon, when seen in perspective, the rigid horizontal base is well-defined as they recede and emerge into a bank of storm cloud. The stifling heat is an evidence of the invisible, transparent storm-cloud overhead, saturated with vapour and shutting in the radiant heat that has passed through it but is powerless to return. Its highly refractive properties bring into view distant objects hitherto invisible. Its low specific gravity reduces the barometric column, and smells and earthy vapours rise out of the ground; animals are uneasy, plants close their flowers, and the winds are hushed. If a puff is generated it whirls along in a little dust tornado, reinforced by the heated dusty column. Although so invisible the storm already exists. The transparent air above is a loaded mine, ready to explode on the slightest disturbance of its equilibrium, like a saturated chemical solution flashing into crystals when one single crystal is dropped into its midst.

A constant rumble by day and distant lightning by night betrays the direction of its approach. At length it rises in solid columns above the horizon. The transparent cloud is transformed into vapour, so dense that a darkness like that of an eclipse sweeps over us as it passes by; a sudden blast of icy wind, brought down from above, slams every door, and the house can scarcely be closed in time to keep out the tempest of hail and rain; the lightning escapes in incessant streams; the prodigious condensation that has set in, and the amount of heat that has been set free, is measured by inches of solid water over half a continent.

The force of the hurricane need excite no surprise, for when the tempest howls among the rigging, and the waves, lashed into fury, sweep the decks from stem to stern, what urges the stately vessel in her onward course? It is only another storm of similar character, but of greater force, confined within the narrow limits of the cylinders. The heated vapours there condensed bring down the massive piston with the force of a hundred tornadoes, and the iron hull, sweeping aside the winds and waves, scorns the opposing efforts of a dissolving cloud.

The following description of the great storm, 1877, will be of interest to my engineering friends.

## THE GREAT STORM OF MAY, 1877.

The great storm which extended over the whole of the Argentine Republic on the 2nd and 3rd of May, 1877, will long be remembered on account of the extraordinary floods which desolated the country, and possesses especial interest for the engineer on account of the severe test to which it exposed all engineering works which intercepted the waters in their descent toward the sea. The storm itself, though of such unprecedented violence, was peculiarly normal and simple in its character, and it was preceded by phenomena which gave unusual timely notice of its advent to the meteorologist. The rainfall, even in the month of March, had been considerable, amounting to 6·15 inches in Buenos Ayres, and to a much larger amount, unfortunately not reliably recorded, in the interior of the province, but evidently increasing in quantity in proportion to the distance from the sea. Immediately after the equinox, as though following the sun across the equator, enormous waves of heated and saturated equatorial air flowed southward, over the whole region of the pampas, condensing at first near the Andes, and gradually, by retrogression, deluging the whole southern area of the continent. These waves were unusually remarkable as regards their periodic regularity throughout the whole month of April, as will be seen by the annexed diagram of the movements of the barometer. They succeeded each other with pendulum-like regularity every five days, each succeeding wave being greater than the last, and invariably accompanied, as each lowest point was reached, by a disturbance or storm of more or less magnitude. Thus:

| | | | |
|---|---|---|---|
| April 2, | bar. fell to | 29.67 | with storm and rain. |
| 7, | ,, | 29·60 | with a pampero. |
| 11, | ,, | 29·58 | with rain. |
| 15, | ,, | 29·54 | with hurricane midnight, and great storm salto. |
| 20, | ,, | 29·58 | with rain and wind. |
| 25, | ,, | 29·49 | with thunder-storm at noon. |
| May 2, | ,, | 29·40 | The great storm itself. |

Assuming the line at 30 inches (fig. 5) to represent the mean height of the barometer, a straight line, A B, making a small angle with this line, touches almost accurately every lowest point reached by the barometer, and indicates very strikingly the enormous accession of heated and saturated equatorial air which was so regularly and gradually accumulating.

The condensation of this prodigious mass of saturated vapour into rain occurred with such rapidity that after the storm the barometer rose from 29·40 to 29·90 almost suddenly on the morning of the 3rd, the temperature of the air as rapidly falling from 70° to 56°. The total rainfall at Buenos Ayres was 8·76 inches in forty-eight hours, of which 6·5 inches fell in the short period of thirteen hours. The storm was accompanied with vivid lightning, thunder, and hail, and a hurricane

**BAROM. AT 9 A.M. REDUCED TO 32?**

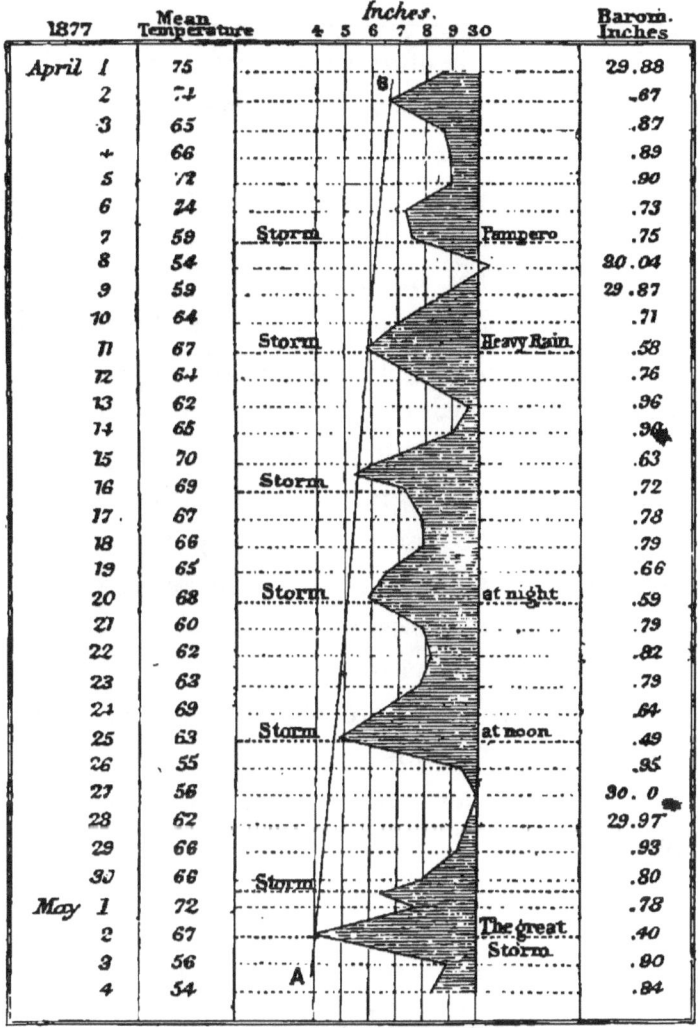

| 1877 | Mean Temperature | | Barom. Inches |
|---|---|---|---|
| April 1 | 75 | | 29.88 |
| 2 | 74 | | .67 |
| 3 | 65 | | .87 |
| 4 | 66 | | .89 |
| 5 | 72 | | .90 |
| 6 | 74 | | .73 |
| 7 | 59 | Storm ... Pampero | .75 |
| 8 | 54 | | 30.04 |
| 9 | 59 | | 29.87 |
| 10 | 64 | | .71 |
| 11 | 67 | Storm ... Heavy Rain | .58 |
| 12 | 64 | | .76 |
| 13 | 62 | | .96 |
| 14 | 65 | | .96 |
| 15 | 70 | | .63 |
| 16 | 69 | Storm | .72 |
| 17 | 67 | | .78 |
| 18 | 66 | | .79 |
| 19 | 65 | | .66 |
| 20 | 68 | Storm ... at night | .59 |
| 21 | 60 | | .79 |
| 22 | 62 | | .82 |
| 23 | 63 | | .79 |
| 24 | 69 | | .64 |
| 25 | 63 | Storm ... at noon | .49 |
| 26 | 55 | | .85 |
| 27 | 56 | | 30.0 |
| 28 | 62 | | 29.97 |
| 29 | 66 | | .93 |
| 30 | 66 | Storm | .80 |
| May 1 | 72 | | .78 |
| 2 | 67 | The great Storm | .40 |
| 3 | 56 | | .80 |
| 4 | 54 | A | .84 |

Fig. 5.

of wind, and, as is often usual, was preceded by a singularly transparent state of the atmosphere, so that the Monte Videan coast, on the other side of the river, upwards of thirty miles distant, rose prominently and clearly into view. It was characterized moreover by other phenomena, of high interest to the meteorologist, but foreign to the object of this paper, and I will only further allude to the extraordinary mechanical energy developed by the mere pressure of the descending rain.

The following diagram (fig. 6) illustrates the oscillations of the barometer during the height of the storm, observed minute by minute, and, as will be seen hereafter, possesses great interest, as they go far to account for the apparently capricious manner in which certain spots are devastated and massive works destroyed while other neighbouring structures, much less solidly constructed, escape uninjured.

The movements of the barometer were so rapid as to be distinctly visible to the eye, and under a lens oscillated with a constant wave-like motion.

Commencing at 10.28 p.m. on the 1st May it will be seen that the barometer rose from 29·60 to 29·79 in one minute, and in one minute more it fell from 29·79 to 29·58. In four minutes more it again rose from 29·58 to 29·81, and thus gradually described the singular curve shown in the diagram.

# BUENOS AYRES.

### BAROMETER OBSERVATIONS TAKEN AT ONE MINUTE INTERVALS IN HUNDREDTHS OF INCHES.

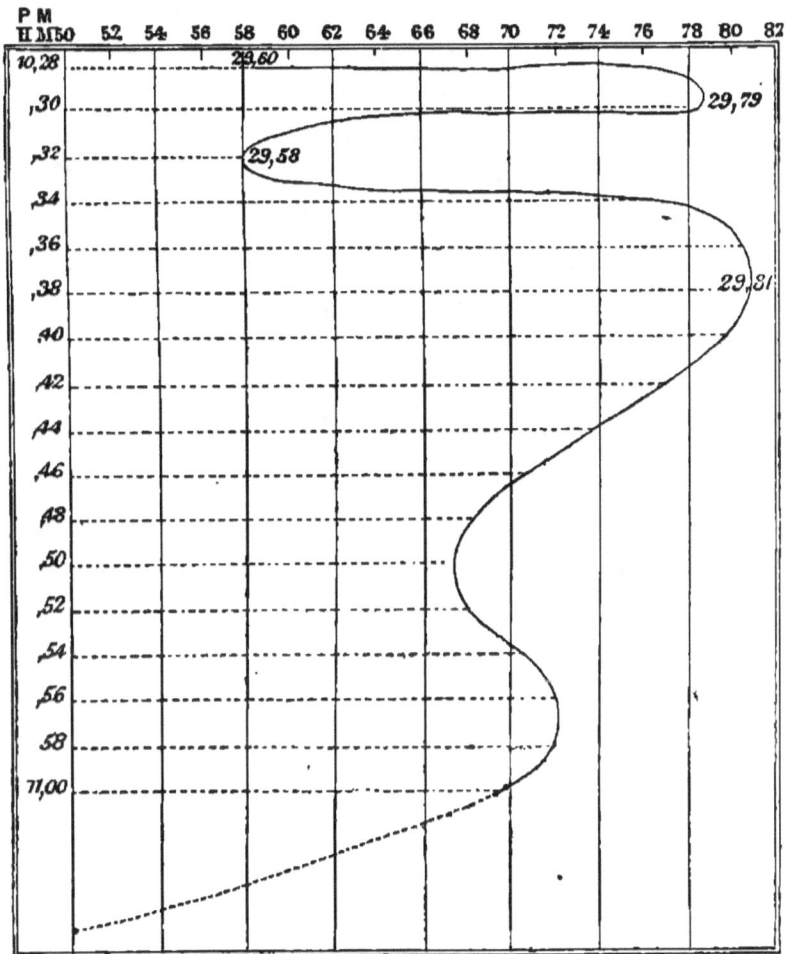

Fig. 6.

The whole of the first rapid rise of 0·19 was simply due to the mechanical force by which the air was compressed by a descending torrent of rain, which at the moment was precipitated like an explosion from a cloud overhead. The rise of 0·2 inch was due to the rebound to restore the vacuum caused by this rapid condensation, and then follows the natural oscillation to restore the equilibrium so violently disturbed. A similar curve was again observed during the storm of the following morning. The electricity set free by this tremendous precipitation was enormous in quantity but moderate in intensity, streaming freely away through this saturated medium in waves and ribbons of brilliant fire. The thunder was consequently continuous in its character, with none of those definite volleys of sound which are usual with electricity at higher tension passing through dry strata of non-conducting air.

The hail-stones which accompanied the lightning were of large dimensions. Their formation in this heated mass of vapour was probably due to the descent of cold air from higher regions into the vacuum produced by condensation.

The low-lying portions of almost every railway in the Republic suffered more or less. The damage done to the Western line was considerable, but the Southern, the North-Western of Monte Video, and

## Subterranean Rivers.

the Campana lines were naturally from their position the greatest sufferers. The water in the whole valley of the Salado continued to rise for many weeks after the termination of the storm, interrupting to a serious extent the traffic on the Western Railway and destroying a large quantity of cattle. The number of sheep destroyed is estimated at 10,000,000. This is to a great extent accounted for by the continuous rains of the previous month, commencing in the west, and gradually saturating the whole country as the storm moved eastward. This prodigious rainfall was thus an addition to a wave of water which was already slowly moving downward from the high lands through the saturated strata into which it had penetrated. It is a singular phenomenon that the whole of the interior of the pampas has no natural drainage in the form of rivers; the rainfall over this vast area gravitates into the pervious soil, until arrested by water holding strata at greater or less depth, and it is more than probable that at these depths, delayed by the friction of the saturated soil, slow underground rivers of drainage move gradually down towards the sea, their course being determined by the configuration of the water-bearing strata beneath. The continuous rise of the water in the low valley of the Salado, so long after the storm had ceased, is probably due to the cropping out of these descending springs.

## Damage to the Campana Railway.

The Campana line, after running 24 kilometres out of Buenos on high ground, traverses the great banya or marsh which forms part of the modern bed of the Parana. It consists of a deep bed of alluvium, now slowly accumulating, partly from the deposit left by floods from the great river on its eastern margin, and partly from the *débris* brought down by storms from the baranca or cliff which bounds the great plains of the pampas and forms the western margin. These great plains slope downwards from all sides with an inclination of eighteen inches per mile near the river, and with steeper and steeper gradients as they recede from it. The water during great storms being but little absorbed, precipitates itself suddenly down these slopes in torrents of destructive dimensions loaded with mud. The alluvial marsh thus formed consists of alternate beds of soft *river* mud and silt, with dark beds of vegetable mould, with occasional layers of river sand. Its depth varies generally from a few feet to 100 feet, and probably much more. It affords good pasturage in the summer months, and is intercepted by a few muddy streams and lakes, and rests on older beds of similar origin, but more or less hardened by the infiltration of lime into a compact rock called *tosca*. The railway embankment crosses this marsh for a distance of forty kilos, passing however over two projecting spurs from the high land above it. It thus completely intercepts three distinct valleys or watersheds, outlets for the water being provided for by numerous large culverts and bridges. The damage done by the storm amounted to the total destruction of some of these culverts and bridges, and the denudation and partial destruction of a considerable length of railway embankment. The author was on the line during the great storm on the 3rd and witnessed the devastation, the banks melting away so rapidly that any delay in retreating with the engine would have rendered retreat impossible. One of the curves illustrating the pressure from the descending rain was made on this occasion. The immediate effect of this enormous rainfall was to convert the level marsh into a vast lake, from six to eight feet deep, and measured by miles in breadth and length, the water pouring over the baranca and down the three valleys alluded to in wide rivers. The ordinary law of water finding its level was disregarded, for the varying barometric pressures shown in the diagram weighed locally on the lake in one spot and removed pressure simultaneously in another, so that huge undulations swayed to and fro like an ocean tide. The wind, which rose to a complete hurricane from the west,

184                    *The Destruction.*

carried these tides forward and brought down intermittent volumes of water with destructive velocity on the banks and bridges. The waters, moreover, penned up by the embankment, were in many places three and four feet higher on the upper side than on the lower, and flowed through the culverts and bridges like water through a milldam, scouring out the alluvial soil into deep trenches. The banks crumbled away before this incessant wash, leaving the rails and sleepers suspended in the air, while in some spots the waves leaped over the line or carried it bodily away. An opening, made purposely to ease the pressure in the banks, was rapidly enlarged to a gap many yards wide. The damage was thus mainly due to the effect of the wind, and would have been far less in its absence, but it was moreover increased by the water-way being insufficient for a volume of water so sudden in its descent and so unexpected in volume. The embankment was more or less undermined everywhere, but more especially where atmospheric waves as described gave local impetus to the waters, or where the soil was of a silty character, easily dissolved or destitute of vegetation. It offered more resistance where thoroughly covered with vegetation, and especially where overgrown by a species of creeping solanaceous plant with fleshy deep roots. It was moreover considerably protected in many places by the enormous layer of vegetable *débris* brought down by the flood and arrested by the bank. It is difficult with the materials at disposal in a country where not a single pebble is to be found to resist the force of waves by any practical engineering device. The only feasible resource would be to destroy them by means of a timber barricade on short piles placed a few feet from the embankment, the planking not extending deep into the water, but simply breaking the surface undulation. Experiments with this object have at the suggestion of the author been made on the Southern Railway with a promising result.

The three valleys intercepted by the railway are the valley of Las Conchas, drained by the river of the same name, the Escobar Valley, and the Lujan Valley, traversed by the River Lujan. It was in the great valley of Las Conchas that the principal damage occurred; the large bridge over the main river both here and at Lujan escaped without injury, but the bridge at Pacheco, consisting of a centre iron girder of forty feet span, with two timber side spans of thirteen feet, each with massive brick piers and abutments, was destroyed, the centre girder and roadway, and the two piers, being carried bodily away to a distance of sixty yards, but fortunately without any damage to the girders. The abutments were also more or less undermined. Many of the culverts in the valley were undermined or destroyed, and the line breached in several places, the banks everywhere being seriously denuded.

In the Escobar Valley, the bridge was an iron bridge of sixty-four feet span with brick abutments. This bridge was swept bodily away, together with one of the abutments. The banks also were considerably damaged by the wash of the waves.

In the Lujan valley the Pescado bridge, consisting of three spans of twenty-five feet, each with iron girders and brick piers, was seriously damaged. One of the piers was carried away, and two of the girders thus fell into the water, but were not swept away; while the Salado bridge, an iron bridge with two spans of forty feet and two of twenty-five feet, was placed in great jeopardy by the sinking, to the extent of sixteen inches, of one of the abutments which carried the large girders.

There was no deficiency of waterway in this valley, the level of the water on both sides of the line varying very slightly on account of the large opening at Lujan bridge, which is 724 feet long.

The immediate engineering problem was the restoration of the line for the resumption of the traffic in the quickest possible manner, and this was done with such rapidity that trains were again regularly running on the 22nd May, or in twenty days after the storm.

With respect to the earthwork, every available resource was put in requisition for the reconstruction of the line, and it fortunately happened that an engine was available for the purpose at the Buenos Ayres end of the line, while another was imprisoned at Lujan. The total quantity of earthwork required for renewal was 63,000 cubic metres.

The temporary restoration of the bridges and culverts was a more anxious problem. The large quantity of timber in store at Campana, and the pile-driving machinery belonging to the contractors, afforded valuable resources for the erection of piled timber structures, so constructed as not to interfere with the future permanent reconstruction.

Whenever the *tosca* was near the surface, this was an easy problem, but on taking soundings at Pacheco and Pescado no bottom was found at fifty feet depth, and so soft was the singular formation at Pacheco that the piles sunk twelve or fifteen inches under a blow of the monkey even at this great depth, while it was moreover so elastic that they rebounded after the blow, and followed the monkey through a still greater space, so that it was only by securing them after each blow that they could be kept down at all.

Ordinary piling was thus impossible, and a novel

device was resorted to, which proved eminently successful.

Large transverse balks of timber were laid on the mud bolted together, but separated by distance pieces, the piles were then driven between these balks with cleats bolted on each side to prevent their passing through, and the frame was then bodily driven into the mud with short falls and a very heavy monkey. In this manner these transverse timbers were forced six or eight feet into this elastic deposit, until it was impossible to drive them permanently farther. The piles were then efficiently cross-braced, and these bridges have proved thoroughly efficient, without the slightest displacement under the heavy traffic to which they have so long been subjected. Where the ground was good the culverts and bridges were restored on ordinary driven piles, while some culverts were entirely filled in, and afterwards reconstructed.

The permanent restoration of all these works was the last step. The water-way in the Las Conchas Valley being insufficient, it was resolved to increase it by the construction of a viaduct, near-the mill at the lowest part of the valley, 300 feet in length, thus practically doubling the total length of all the existing openings, including the bridge over the river.

The length was arrived at from the observed differences of level on either side of the line at the maximum height of the flood, and is calculated to reduce the velocity of the outfall within limits that will not disturb the grass and vegetation beneath the viaduct. The soil here is a dark alluvium of considerable toughness and elasticity, and the viaduct consists of a series of hard wood trestles, strongly cross-braced, dug four or five feet into the marsh, each trestle resting on a bed of cement concrete three feet wide. The trestles are fifteen feet apart, and carry a continuous double line of iron girders, on which the cross sleepers and rails repose. The expansion and contraction, amounting to two inches, takes place by the slight yield of the frames in the elastic soil. The cost of a bridge or viaduct is always a minimum, when the span is so taken that the cost of a pier is exactly equal to the cost of the girders, and the length of span was then determined, the total cost was £3 sterling per foot run. This viaduct has been some time in use, and forms an efficient and durable structure. At each extremity the line is protected by sheet piling.

PERMANENT RECONSTRUCTION OF THE BRIDGES.

On further examination of the river bed at Pacheco, where the bridge had been temporarily replaced by a timber structure on sunken frames driven into the mud, as already described, it was found that at a depth of fifty-two feet a solid foundation could be obtained on the *tosca*; the permanent bridge has therefore been built with spliced hard timber piles driven to the solid rock.

At Pescado, however, even at a depth of sixty-five feet, no bottom was found; the ground in fact became worse, and at this great depth a quicksand was reached, into which piles sank two feet with a blow. The original abutments and piers had been built here, as in other cases, on a hard superficial bed or layer of sand found at the surface, but of no great depth nor extent. The destruction of the bridge arose from the undermining of a portion of this stratum. Immediately beneath this bed of sand, to a depth of twelve or fourteen feet, there was found, moreover, a layer of hardened alluvium, offering very considerable resistance to the driving of piles, which only sank easily after passing through the hard layer.

Experiments were made with the object of utilizing this layer of hard soil, and were highly satisfactory. A pile of the following description was found to offer so much resistance, that the pile itself was destroyed by the repeated blows of the monkey before it had passed through this stratum of hard soil. The piles are of hard wood, with an enlarged inverted pyramid at the base, as shown in the sketch. The base is formed by simply bolting four large timber wedges to the sides of the pile. This modification has avoided all disturbance of the superficial hard layer which overlies the quicksand, and affords a foundation which could in no other way be obtained.

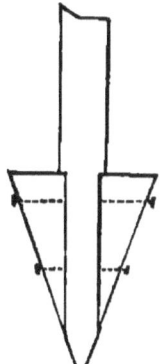

CONCLUSIONS.

The disaster has afforded experience of much value. Massive brick piers and heavy abutments were out of place on such foundations. When built on piles they remained secure, although entirely undermined between the piles. In this case the undermined portion has been filled with rammed cement concrete, and means taken to prevent further excavation. At Escobar the massive pier, weighing thirty-two tons, was built in brick and cement of such solidity that the whole pier, including the concrete foundation, was swept bodily away to a distance of 250 yards, without any fracture of the pier or the disturbance of a single brick. The eroding action of water, rushing through a bridge or culvert, increases with the depth as well as the velocity. In the centre of an opening on a level bed, the descending current surface forms an inclined plane from the upper to the lower level, the depth decreasing uniformly; but the velocity increases in a higher ratio until arrested by spreading away on all sides into the lower water. The scouring power is thus greatest on the lower side of the culvert, and a little beyond this point, where the current dies out, the eroded material accumulates and forms a bank or island.

Again, on the upper side, the water which flows freely away at the centre banks up against the abutment and maintains its full depth, while the current in its passage below converts the base of this apparently quiet column into a bottom whirlpool, which acts with destructive force on soft soil, frequently scooping it out into dells and hollows alongside the wing walls of the abutments.

There is another scouring effect, due to depth, of considerable importance. The excavations made for the foundations being only loosely filled in, the scour commences near the abutments, and with increasing force as the depth increases, while at the slight depth which occurs at the middle of the span the turf is not even disturbed. In all cases the excavation should be securely filled with the best material at hand, well rammed. In our case clay is successfully used, but after every storm the ground should be invariably restored as near as practicable to a level bed.

In all cases where the piers consisted of piles, whether of timber or iron, they suffered no damage. To avoid the numerous difficulties attending the use of masonry on such bad foundation, piles have been exclusively substituted for brickwork even in the abutments; a little extra length of girder allows the abutment piles to be driven into the bank itself, the toe of the earthwork with its natural slope being protected by timber sheet piling. Even on the recurrence of floods of greater extent than the last, all that could now happen would be the sweeping away of the toe of these embankments until the waterway had sufficiently enlarged itself, without any disturbance of the bridge. The restoration would thus consist simply of a few yards of earthwork, which can be rapidly restored. The timber viaduct, before described, is in every respect designed strictly in accordance with these principles, which are the result of experience, and which it is confidently believed are worthy the serious attention of every engineer who may have similar difficulties to overcome.

Since this paper was written, these new works have been put to the test. Extensive floods, almost equal to those described, but without much wind, have again been experienced, without any damage whatever to the line.

### Storms at Buenos Ayres.

The meteorologist will forgive me if I introduce two or three further examples as illustrating the character of those great disturbances. After several very oppressive and unbearable days, with constant equatorial wind, a little cloud, "no bigger than a man's hand," appeared in the south on the evening of the 27th June, with distant lightning, and we knew the southern winds were gaining ground; it, however, soon disappeared. On the following evening, at the same hour, clouds again collected in the south, this time in dense masses, with lightning, but after covering half the sky, they all again retired during the night. On the third evening, the clouds again rose up from the south in dense and massive columns, and soon entirely covered the sky, with heavy rain and lightning; the thunder was constant and deafening, and the lightning streamed in ribbons of fire among the clouds, but the atmosphere retained its temperature of 60°, after the storm had passed, and we knew at once that it was only a lull; on the following morning, the storm again returned with far greater magnificence. The lightning was constant and terrible; it poured down from apparently fixed points in the sky in vertical vivid streams 45° in length, as though a reservoir of molten metal had burst overhead. It was almost entirely free from the usual zigzag form. The clouds drifted in vortices in all directions, and the thunder was absolutely continuous. A dense darkness added to the effect, and the rain deluged the town, torrents of water of great depth pouring down into the river through the narrow streets. The storm ceased with us at noon. A friend who came from Salto, 250 miles up the Uruguay, informed me it had commenced there at 3 a.m. on the previous day, and he had travelled the whole distance through the storm. Moreover, the Uruguay had been rising during the last eight days. The rain must therefore have prevailed far north of Salto.

The area over which this great storm prevailed was enormous, and it is difficult to realize the prodigious mass of vapour that must have been condensed, and the corresponding amount of vapour and electricity set free. These storms are invariably followed by cold refreshing winds. Although in this instance the atmosphere retained its temperature of 60° for many hours, the south-west wind ultimately prevailed, and drove away the hazy cloud in mottled masses. The temperature fell to 50°, and in the night to 43°, and the storm was followed by a long spell of bright, cold, lovely weather. The storm came to us from the south in successive invasions, beginning with the little cloud, with lightning in the south on the evening of the first day, and the re-appearance of this phenomena on successive evenings in a larger and larger scale. Storms had, therefore, probably occurred to the south of us. The diurnal advance and retrogression was probably due to the accession of heat from the sun during the day, which reinforced temporarily the equatorial or northern wind against which they were advancing. The violence and duration and the excessive rainfall were due to the fact that the point B (fig. 4, p. 172) remained for some time nearly stationary, the motion of the whole system towards the south being neutralized by the advance of the tongue of cold air, so that a constant rapid river of vapour was precipitated at one shot.

The varying violence of these storms depends on the amount of local reinforcement they meet with in their course. No stronger proof of this can be required than the fact that the disturbance always advances bodily, regardless of the wind, in the direction determined by condensation, that is, from drier regions towards regions of greater moisture.

After this storm continual phases of heat and cold occurred, with storms on the 10th and 24th of August, until we come to the Santa Rosa storm of the 30th. The boisterous weather that generally occurs about this period in the Argentine republic has given Santa Rosa a very unenviable notoriety, and doubtless some of the largest and most destructive floods on record have occurred on or about the anniversary of this turbulent saint. I was anxious to ascertain the grounds of this general belief. Previous to the 28th, the oscillations in the barometer curve were very large, and, as is generally the case, the corresponding intervals between the vertices very small. The converse is also generally true, so that similar waves give similar intervals. The changes in the weather are thus periodical, in epochs varying from a few hours, as in the present case, to seven or eight days.

The heat was again considerable; large areas of underwood and dried vegetation were on fire over many square miles among the islands, and dense masses of smoke obscured the atmosphere and increased the heat. The 28th was a dull, heavy day, clouds from the north-east, charged with moisture, collecting into heavy horizontal strata with straight lines. On the 29th we had a storm which lasted all day, with incessant thunder and lightning, but there were intervals of bright sky, as though the clouds were projected spasmodically and not in a continuous stream. The temperature was 60°. The storm travelled to and fro as the junction of the opposing currents changed its position. The lightning was extremely brilliant; ribbons or arcs of great width, and from thirty to forty degrees in length, streamed upwards from the river, disappearing abruptly behind a mass of dense cloud. It was not zig-zag in form, but unevenly wavy and ribbon-like, sometimes in the form of descending arches not reaching the ground, and sometimes suddenly exploding in the air in the form of the letter S.

The forms assumed by the lightning are consistent with the explanation which I have attempted; masses of vapour condensed on the earth's surface not only disturb the electrical equilibrium by the setting free of their own electricity, but they isolate the overcharged layers above them, and the discharges may partly owe their streaming form to the partial vacua that are produced.

The rain was enormous in quantity, and though the heaviest cloud formation was in the south-west, detached clouds floated by at low elevation with great velocity from the north-east. The cloud advanced steadily against the wind, partly from new additions, and partly from condensation; the discharges of electricity set free were confined to the mass of cloud that was in more immediate contact with the earth.

About 10 p.m. the storm was at its height, and the spectacle was magnificent. Intensified by a temporary hurricane from the south, with a deluge of hail, the contest still continued. Distant thunder was incessant. "All day long the noise of battle rolled among the hills." At 9 a.m. on the morning of 'the 30th, St. Rosa's day, the tempest broke out again with still greater fury; a south-west hurricane, with destructive hail, shook our building to its foundation. The tiles were stripped from our stations, and much damage was done. The rain still fell in torrents for many hours, clouds were still drifting from the north-east and east as though unwilling to abandon the contest. The storm, however, disappeared finally with mizzling rain and fog on the morning of the 31st, when the barometer suddenly rose from 29·65 to 30·10 inch. The area over which this storm extended was again enormous. It prevailed over the whole country, from the Cape to the Pacific, and floods were very general. At Monte Video 5¼ inches of rain fell in forty hours. In all cases the numerous windows that were broken faced the south.

In addition to these normal storms, the contest often locally assumes the form of a tornado. On the 30th September, the evening was clear and bright as usual, but I observed an extraordinary fall in the barometer. We were awoke about 4 a.m. by a tempest of rain, with violent thunder and lightning, which appeared to move all around us. The following morning was hot and oppressive, the air was still, with a peculiar haze which, no doubt, was the cause of the intense heat, as, although the sun's direct rays at high tension pass through such a haze of vapour, they are powerless to return in the form of radiant heat, and accumulate as they would in a greenhouse. I was called to Campana on business, and at sunset my attention was directed to a singularly dense and perfectly stationary bank of dark cloud in the south-west. The wind was north, and very light. The edges were sharply defined, and behind it there played a perpetual fountain of quiet summer lightning. I watched it for a long time, but it retained its motionless appearance. At 9 p.m. it rapidly changed its aspect; it became terribly dark, so that the full moon was obscured, and in an instant it rushed towards us, accompanied with clouds of dust and a hurricane of wind so strong that it burst open the large heavy doors of the station, and it required our united strength to preserve our own door from the same fate. It hurled down trees and tiles all around, and filled the house with dust and sand. After a partial lull at about 10 p.m., the wind changed instantly from north to south; the thermometer fell in ten minutes from 86° to 58°, and now we were exposed to a cold hurricane of hail and rain, which forced its way through the doors and windows and flooded the cottage, converting the previous dust into mud, which covered the house and windows. Anxious as we were to watch the progress, it was impracticable to get outside during such a shower of hail and tiles. The noise of the storm prevented our hearing any thunder, if there was any, but lightning of the most brilliant and singularly diffused kind was so incessant as to light up the whole neighbourhood. Later on, we were startled by half-a-dozen rapidly-repeated volleys of thunder. It would thus appear that a tornado, or small cyclone, had passed over us from west to east.

During the northern wind we were in its eastern periphery, and during the southern wind we were on its western periphery. The stationary cloud was probably the cloud of dust drawn up in the vortex, and as the change to a southerly wind was permanent, it is evident that this disturbance arose on the confines of the northern and southern currents.

The barometer fell to 29·33 inches, the lowest record we have registered in South America. This storm extended also over a large area. The same phenomena occurred fifty miles to the south, at Buenos Ayres, at nearly the same hour. At 1 p.m. there was a storm at Paysandu, 200 miles to the north. The river fell rapidly, before the north-east wind, but rose suddenly the following day with the change, and flooded the low grounds. The barometer rose, as usual, very suddenly after the storm. Many vessels were driven ashore in the river, and several houses were wrecked; and, unfortunately, my own registering thermometers were carried away.

The general explanation of these disturbances leads to the conclusion that the accessions of heated air from the equator arrive in successive rolling cylinders, or in *vertical* cyclones, these cylinders having at first the form of long ellipses, but gradually contracting by the condensation until they are absorbed.

A succession of such rolling ellipses explains the contrary current so invariably present overhead, the fall and subsequent rapid rise of the barometer, the haze and heat preceding the storms, and the singularly regular periodicity, and other observed phenomena otherwise inexplicable.

Except during these short intervals of storm, the climate of Buenos Ayres is, as a rule, magnificent. The heat and moisture during the prevalence of northerly winds, although very disagreeable, seldom interfere with the perpetual sunshine and clear blue sky. The moisture in the houses, which so disagreeably shows itself in the wardrobe in the form of mildew, is due, not so much to the actual humidity, as to the condensation which takes place on cold objects during the very rapid changes of temperature and humidity which are so frequent. The effect of the river on the humidity is, however, very considerable.

In reviewing my observations, the danger of drawing conclusions from partial data is illustrated by the following singular coincidences. It is established, beyond a doubt, that there is little, if any, connection between the moon's changes and changes in the weather. But the following table is a list of all the great barometer depressions between September and December, 1876.

I have added the dates of the nearest full moons, which certainly coincide in the most remarkable manner, but I could trace no such connection in previous or subsequent storms, although a constant tendency to periodic disturbance every seven days is very manifest.

| Barom. | | | | | |
|---|---|---|---|---|---|
| Oct. 2. | 29·35. | Dust Storm Buenos Ayres. | Full moon, | | Oct. 3. |
| Nov. 2. | 29·40. | Storm Rosario. | ,, | ,, | Nov. 1. |
| Dec. 2. | 29·40. | Great dust storm. | ,, | ,, | Dec. 1. |
| Dec. 30. | 29·43. | Dust storm. | ,, | ,, | Dec. 30. |

In a country where cloud is so unfrequent, one of the phenomena which at once excites the attention of a stranger is the magnificence of the sunsets. The gorgeous masses of gilded cumulus appear to be brought out for the occasion every night, and to be immediately removed as soon as the stately ceremony is ended. This effect is probably due to the fact that, during the heat of the day the air, owing to its higher temperature near the heated surface of the earth, absorbs a larger amount of moisture. As the setting sun retreats, a zone of the western horizon, deprived of its rays, is suddenly chilled, and the vapour is temporarily converted into the sunset clouds which exist for a certain time, until equilibrium is restored, when they are again dissolved; a constant volume of cloud thus follows the setting sun and marks the boundaries of night and day. The gorgeous colours depend on the obliquity of illumination and the decomposition of the light through these vast prisons of air and vapour of varying density. The most refrangible rays, the purple and violet are the last that are seen, and close the glorious pageant.

## CHAPTER XII.

*THE URUGUAY AND SALTO.*

MATTERS of business connected with the North-Western railway of Monte Video led to a lengthened residence in the northern part of Uruguay, and afforded one of the most interesting episodes in my travels. I had long looked forward to the opportunity of ascending these great rivers, and in spite of the attractions of Buenos Ayres we gladly welcomed some nearer approach to the tropics.

The Uruguay runs nearly parallel with the Parana, enclosing a true Mesopotamia or "Entre Rios" between the rivers. Branching off to the east in latitude 27°, it drains the mountain ranges of the southern tropical provinces of Brazil. It is a magnificent river, in the rainy seasons rivalling the Parana in its actual volume of water, and more than doubling it in velocity. Instead of flowing through a level bed of alluvium, it has cut a steeper channel through hard and rocky strata, and its floods subside with great rapidity. Ocean-going vessels, however, ascend at all times as far as Paysandu, about 140 miles from the Plate, and except during long periods of drought the large river steamers run through to Salto, which is 215 miles from the Plate. Beyond Salto, for a distance of 100 miles, as far as Santa Rosa, the rocky bed of the river is steep and irregular, giving rise to a series of rapids and falls which intercept the navigation. Beyond Santa Rosa it is again navigable, and is the channel of an extensive carrying trade. It forms the only outlet for the valuable produce of the rich and fertile southern provinces of Brazil. The railway from Salto to Santa Rosa was thus a commercial necessity, and cannot fail to command a large trade when completed. At the great falls above Salto the river is nearly a mile and a quarter wide, with a leap at low water of 15 feet over a succession of diagonal reefs of primary rock. The total fall is 25 feet. During high floods, however, this rocky bar is totally submerged, and powerful steamers occasionally ascend the rapids through a rocky channel on the right bank of the river. The river at Salto rises no less than 50 feet during these great floods. The large and powerful steamers which ply between Monte Video and Salto, calling at Buenos Ayres *en route*, are fitted up in American

## The Steamers. 207

style, with every luxury; an excellent dinner is served on board in the large and handsome saloon, accommodating from 150 to 200 guests. The cabins or sleeping berths are large and elegant, furnished with sofas and good beds and every convenience. The fare, £4 5s., is high, but the journey in one of these floating hotels is a luxury of which we can form no idea in England. It is of course greatly enhanced by the lovely climate and the wild and wooded scenery of the islands and banks. The passenger traffic is very large. The only drawback to a landsman is the rough water which in bad weather is sometimes encountered on the Plate. Leaving Buenos Ayres at 5 p.m., we arrive at Salto about 4 p.m. the following day. When the river is low the large steamer is left at Paysandu, and the journey is completed in smaller boats. On leaving the Plate, after passing the fortified island of Martin Garcia we enter the estuary of the Uruguay, which may be regarded as a continuation of the great estuary of the Plate, and extends as far as Fray Bentos. At Hugueritas several branches of the Parana unite with the Uruguay, and form a communication between the two rivers, and a few miles above Hugueritas the estuary assumes the form of a large lake of brilliant clear water, nearly sixty miles long by five or six miles wide. This lake is without islands, and is generally shallow, with a deep channel half a mile wide throughout its length.

On leaving this lake near Fray Bentos we sail among numerous large wooded islands forming a delta thirty-five miles long and four or five miles wide, very similar to that of the Parana, terminating at Paysandu. The river proper thus commences at Paysandu, 150 miles from the Plate, and though occasionally very wide it is constantly confined within a rocky channel throughout its course to Santa Rosa. That the whole of the matter brought down during the floods of this great river is employed in filling up the delta is evident from the constant transparency of the waters of the great lake below.

.The large Liebig Extract establishment is a prominent object on the Tosca Baranca at Fray Bentos, and as we ascend higher up the river the right bank is composed of cliffs of rock 100 feet high, with extensive forests of palms and small trees, and the rocky shores are everywhere covered with underwood and creepers and rich vegetation. The Yatai palms appear to thrive in these sandy wastes, but all the larger timber has been cut down by the charcoal burners, whose destructive presence is often indicated by columns of curling smoke among the foliage. The waters of the Uruguay increase in temperature as we proceed north, but not with the same regularity as those of the Parana. In passing through the mirror-like lake we noticed that the smooth surface of the undulations caused by the steamer resembled watered satin,—a beautiful effect, due to caustics formed by the brilliant reflection.

The little town of Salto has been in former days a trading town of great importance. Ill-natured critics maintain that its wealth was mainly derived from the extensive smuggling trade which its proximity to the frontiers of Brazil and Entre Rios would doubtless tend to encourage; but its situation as the terminus of the navigable portion of the Uruguay must always ensure a considerable trade of a legitimate character. Its present depression, from which it is fast recovering, is doubtless due to the incessant revolutions and political plunder to which this little republic has so long been a prey. It is a grievous calamity in these small Spanish republics that the political power has been retained by a few native families who only regard the administration of the country as a profitable occupation, and sacrifice, without hesitation, all its highest interests to personal benefit and the acquisition or maintenance of wealth and political power. This grievous canker has unfortunately been fostered by independent circumstances. Nations, like individuals, cannot escape the stern moral law that the spasmodic acquisition of wealth results invariably in demoralization and ruin when obtained without labour or industry. The tribute paid to ancient Rome undermined and destroyed the empire far more effectually than any external enemy would have done. Spain fell rapidly from the height of prosperity to the lowest state of helpless misrule under the influence of the unearned wealth which flowed in after the discovery of America. Venice and the Italian republics rapidly succumbed under similar baneful influences, and it is a question whether the enormous tribute paid by France to Germany has not proved more fatal than any material destruction that could have ensued from the expenditure of the same amount in offensive warfare. The present condition of Turkey and Peru and other states only too effectually confirms this invariable law.

In the case of Uruguay the evil commenced with the Paraguayan war, when large subsidies of Brazilian gold were absorbed by the needy and unscrupulous adventurers who ruled the country, debauching alike the governors and governed, and destroying every vestige of patriotism, honour, and honesty. Large amounts of money brought into the country by settlers and emigrants were robbed or wasted by one side or the other, during incessant revolutions got up with no other object. The railways, promoted by reckless guarantees, introduced large amounts of capital; and lastly the public loans followed, and completed the moral and material ruin of both the plunderers and the plundered, and left the former without either the will or the means of meeting the engagements incurred. The whole of this wealth, instead of being invested in productive works, was absorbed in profligacy and personal extravagance; the rows of costly Gothic, Moorish, and Byzantine quintas which cover the suburbs of Monte Video, now going to ruin for want of the mere means of maintenance, furnish an eloquent memorial of a period of public vice and scandal which in private life would have been corrected by the prison or the treadmill. It it is thus but too sadly evident that both politically and morally the sins of the father are visited on the children, and that grievous administrative mismanagement is still more or less the inheritance of every spot where the race retains political power. In a country where senators or representatives cannot be found with sufficient patriotism to refrain from voting themselves large personal incomes for their services, it is evident that, with all its disadvantages and serious evils, a firm and honest dictatorship is a boon which a nation would be very loth to yield up; and it is undeniable that the rapid progress of the country under the rule of the present governor is most remarkable. Encouraged by the peace and security which is now resolutely and firmly maintained, emigrants are again returning to cultivate the great fertile fields so long neglected. The herds of cattle and horses, which had been entirely destroyed, are again rapidly covering the pasture lands. A disciplined military force, armed with Remmington rifles, has, it is hoped, for ever put an end to the time when, under protest of being revolutionists or government troops, as best suited their purpose, hordes of savage Gauchos, armed with a bamboo lance, and led by some brigand adventurer, galloped unresisted over the country, levying contributions and sweeping away before them the remnants of horses and cattle which the helpless settlers had not had time to dispose of in Entre Rios or Brazil. The legitimate resources of the country keep pace with its increasing trade and production. The land is being everywhere enclosed and recovering its value, and buildings are again rising in every direction, not indeed in the form of suburban quintas and pleasure gardens, but in the more substantial and productive guise of warehouses, farm-buildings, manufactories, railway extensions, and public works. As in every other district, the devastated country round Salto and the town itself is participating in the benefits of this general renovation, and cannot fail to go on improving so long as the reins of government are in firm hands, and its resources are kept beyond the reach of incompetent adventurers, who claim a vested right in their maladministration. There is not the slightest doubt that with the completion of the railway to Santa Rosa and its extension in Brazil to Uruguayana, for which a concession with a guarantee has been granted by the Brazilian government, the fortunes of Salto must rapidly recover, and if the line is continued south to Paysandu, where the river is practically a continuation of the estuary of the Plate, and accessible for ocean-going vessels, it will become the natural high road to Brazil, and cannot fail to prove one of the most lucrative lines in South America. The steep and precipitous streets of Salto are in sad repair, and often nearly impassable; and although there are many excellent houses, built with unusual elegance and taste, the evidences of general decay and neglect are but too evident everywhere. There is an excellent and handsome hotel in the town, which would do credit to any European capital. A timber pier runs some distance out into the rocky bed of the river, but the enormous floods, which rise sometimes fifty feet, leave it at times nearly high and dry, and at other times totally submerged.

The terminus of the East Argentine Railway at

Concordia is visible on the opposite high bank of the river. The station of the North-Western Railway of Monte Video is on high ground at the back of the town. The two railways run nearly parallel on opposite sides of the river, and the two termini at the other end are again nearly opposite each other; the line on the other side of the river is well constructed, with excellent stations, and runs through a picturesque but totally barren country, intercepted by marshes and liable to destructive floods. The object in view in its construction is totally incomprehensible, as the natural traffic is solely on the Uruguayan side, and it is impossible that it can ever cover its working expenses after the completion of the Salto line. The guarantee given by the Argentine government on the amount expended cannot fail for many years to remain totally unproductive, and to form one of its most serious unproductive financial liabilities.

My stay at Salto was very short, as it was my object to take up my permanent quarters at Arapey, which is near the present terminus of the railway, and about forty-five miles north of Salto. The line, after leaving the dusty-looking gardens in the suburbs, with their fences of prickly aloes, runs all the way over a camp, which is absolutely monotonous. The heat and dryness of the climate parch up the coarse vegetation, and to add to the neglected and wild aspect, the ground is ploughed up in deep furrows by the torrents formed by the tropical rains that often prevail. The camp has thus none of the verdant appearance that it presents at Buenos Ayres. It is thrown up, moreover, into large undulations, which has necessitated numberless curves in the line, with gradients of considerable steepness. The storms have bared the smooth red sandstone rocks on the summit of many of these elevations, and storm water courses have ploughed up deep rocky ditches in the lower levels. The ballast and embankments consist largely of sands, broken rock, and pebbles, and we meet nowhere with the dark and fertile beds of alluvium found in the Argentine plains. It is difficult to conceive how cattle thrive on such a dry and sandy-looking desert; nevertheless, large herds of cattle are dispersed all over the plain, with occasional troops of wild ostriches or emus, that rush away from the approaching train.

The Estancias are far apart, and we travel miles without the sight of a single human dwelling or even human being, and at the first aspect it seems totally incredible that so large and steady a trade as at present exists could be derived from a temporary station which, at the present time, is situated in the very centre of this vast, uninhabited, and inhospitable wild. The present terminus is at

Santa Rosa, about five miles beyond the Arapey, and consists solely of the wooden station and a few huts that have clustered round it. It resembles in the distance a caravan halting in the desert, surrounded, as it generally is, by the bullock carts, which bring goods a distance of 50 to 150 miles from the surrounding camp and from Brazil, and this through an excessively rough country, without roads or villages or halting stations.

## CHAPTER XIII.

### *THE COTTAGE AT ARAPEY.*

THE only large work on the Salto line is the great bridge over the Arapey, which is approached by a long viaduct, and consists of three spans of 150 feet each over the river, supported on large iron columns filled with cement. The bridge crosses the river proper at an altitude of forty-five feet above the ordinary water; the deep alluvium beds which form the northern bank are crossed by a viaduct of wrought-iron girders on cast-iron piles. This marshy dell is covered by deep water during the great floods, which not unfrequently rise thirty or forty feet in a few hours, and convert this rocky river into a torrent that can be heard for miles. The girders are ordinary trellis-work wrought-iron girders, with a single roadway between them on transverse girders resting on the lower flange. The abutments are built of sandstone, and terminate a high bank on both sides of the river.

During the construction of this bridge a little wooden cottage had been erected by the engineers. It was close to the line and about 250 yards from the river on its southern bank. It stands on the bare hard rock about 150 feet above the river, commanding lovely views of the river valley. It was roofed with corrugated iron, with a second roof of timber to keep out the heat, and has a light timber look-out tower rising above the building. This was the lonely and lovely spot selected by my wife and myself for a three months' residence. It is hopeless to attempt to describe the pure enjoyment afforded by this leisurely examination of a subtropical river valley of exceptional interest and beauty, and the undisturbed companionship with nature which such a retreat afforded. The lovely climate, the glorious star-lit nights, the treasures of natural wonders which rewarded every ramble, and the peaceful evenings spent in their investigation, can never be forgotten. How few are aware that such pure and simple streams of happiness are within their reach!

Our nearest town was Salto, forty-five miles away. The only visible habitation was the ferryman's establishment, on the river about two miles below. We were entirely dependent on the morning train for the basket which brought our simple daily fare; the owner of an estancia several miles away would occasionally send us a sheep, and supplied us with a milch cow, which he changed from time to time, and we kept a numerous family of poultry around the cottage. There was no difficulty with the gun at any time in killing a partridge, or snipe, or wild duck, and other game, but the fish in the river refused all temptation.

We had an old leaky boat for river investigation. A Spanish workman acted as cook, while an Italian who was in charge of the cottage was our general servant, and constant companion in our rambles. This was our whole establishment, although the rocks and caves of the Monte near us were the resort of malefactors of the blackest dye, who on two occasions during our residence were formally hunted out by bodies of mounted police. The only property of any value we possessed were our guns, as we were beyond the reach of any aspiring philosopher who might have taken a fancy to the thermometers or theodolite or microscope. Adjoining the cottage was a "remada" or shelter, constructed of hard wood uprights well secured in the rock, covered with horizontal rafters and a deep flat layer of boughs of the lucuma, which with its thick persistent foliage formed an impenetrable shelter for cattle against the terrible noon-day sun, and was always occupied at night by about a dozen horses that had been left at the cottage and belonged to us, and that no one would accept. The poultry invariably roosted on this shed, which protected them from wandering beasts of prey. A couple of powerful dogs slept under the verandah that shaded the front of the cottage, and frequently disturbed us when a wandering ostrich or cattle approached too near our homestead. A friend from Salto occasionally brought his gun and stayed a few days, and sometimes the station master rode over from the station, which was two miles off, with a letter or a message. We were moreover reminded daily of the outer world by the passing train, as it thundered across the lofty bridge on its journey to and fro.

After selecting a site and fixing my thermometer, I constructed my wind-vane, which as usual consisted of an eagle's feather thrust through a cork,

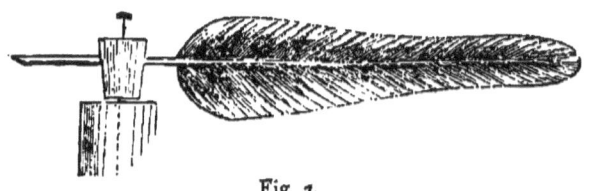

Fig. 7.

with a hole bored at right angles through the quill and cork to admit the long French nail that formed the axis.

On making some experiments with these feathers I was struck with the singular affinity with which

## The Azimuth Table.

wind clings to the vane of a feather as compared with its action on a tin plate or other substance, a specific property of considerable importance in the act of flying. This very efficient instrument was mounted on the flagstaff on the tower. My next care was to establish a permanent site on the rock for my theodolite, and to set up a distant meridian mark. I observed the time by morning and evening altitudes of the sun. The latitude of the cottage is 30° 50′ south, and the longitude 3 hrs. 50 min. 5 sec. west.

It is always my habit whenever I make any lengthened residence to construct what I call an azimuth table, and it has proved so useful to me for rough measurements of all kinds that I will describe in detail the table I left behind me in the cottage at Arapey.

Having determined on a good position, in this case under a verandah facing north, I erect at a convenient height a small table of smooth white pine about twenty inches square. The table is so fixed that the edge lies east and west.

A wire or strong string is then suspended from a fixed support four or five feet overhead, which, passing through a hole at P (fig. 8.) hangs vertically with a heavy weight at the lower end below the table (a bottle of water with a string through the cork is a very good makeshift); a small ball of cork slides up and down the string by means of a slit in the cork (B), a small moveable weight (see C fig. 9)

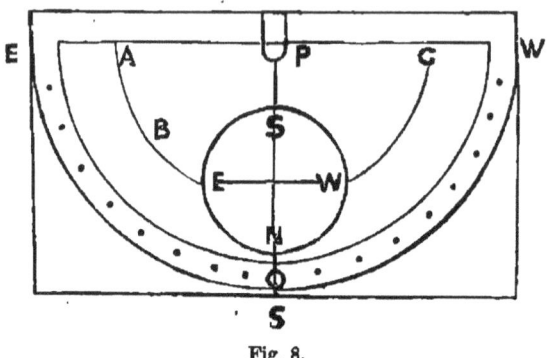

Fig. 8.

carrying a large-headed pin placed anywhere in the circle (S) in one of the sights. My first operation is to draw a correct meridian line (P S) on the clean deal surface. If the time is accurately known, this is at once performed by observing the shadow of the plumb line at the instant of true noon.

If mean time is not accurately known, a semicircle (A B C) may be drawn round the centre (P), and by sliding the ball up the wire we can at any time before noon cause the shadow to coincide with the semicircle and mark its position. If without disturbing the ball we now wait until the shadow again coincides with the semicircle in the afternoon, true south will be half way between these two points in the semicircle, and true noon will be very approximately the time shown by the watch

# The Azimuth Table. 223

at the intermediate time between the two observations.

The pole star furnishes a ready means of obtaining a meridian. It will be due north when our plumb line intercepts E (polaris) at the same time that it covers the star. The meridian line (P S) being thus obtained, the line A C is drawn at right angles through P, and the measuring circle (E S W) is divided into degrees, figured right and left from S.

We are now in a position to properly fix the position in azimuth of any object such as a meteor, or cloud, or sunset, or distant hill, by simply bringing the movable index weight and the movable sliding weight into a line with the given object, and reading the azimuth on the divided circle. But more than this, if we measure the distance A B (fig. 9) we can ascertain the height in arc of the object (S) from any table of tangents, since A B will be the tangent of the angle at C to radius A C, or a rod with a scale of tangents drawn on it may be prepared from which the angle C will be directly read off.

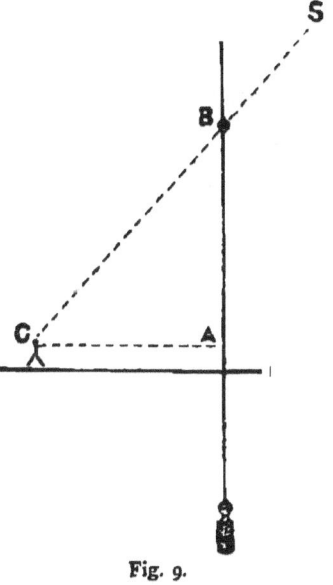

Fig. 9.

A great deal may be done with this simple instrument.

1. The motion of clouds and storms is observed with great facility.

2. The name of any bright star or planet may be determined in its passage over the meridian, as we ascertain at that time both its right ascension and declination.

3. The variation of the compass is conveniently tested.

4. Time at noon is approximately found.

5. The position of the celestial equator and poles is clearly defined by raising the ball to the co-latitude of the observer.

6. The altitude of the sun and moon at noon, and the azimuth of the sun or moon at rising or setting, and their position on the horizon at that moment, are always interesting observations.

7. With a celestial globe and a very little spherical trigonometry, extra meridianal observation is very simple, and a good telescope, steadily mounted, completes a very serviceable rough observatory.

My table contained a solar black bulb vacuum thermometer for solar radiation; a half second's pendulum slung beneath it, with the upper end slightly projecting through a slit in the table, so as to be felt by the finger in the dark, afforded convenient means for estimating the duration of any phenomena such as the flight of meteors, passage of clouds, distance of thunder, etc.

## THE RIVER ARAPEY.

The River Arapey is one of the numerous feeders of the Uruguay; its rapid fall from the high rocky hills in the northern province of Uruguay gives it more the character of an enormous mountain torrent than of a steady river. It has ploughed out for itself a deep chasm in the hard red sandstone rocks of the district, and is cut up into successive reaches by harder strata, which intercept its course, producing a series of falls. These reaches in dry seasons become shallow lakes, with little flow, and the hard rocky bottom forms islands and shallows, which render navigation impossible except during the great floods. The banks are steep and precipitous where the stream is undermining the rock, while the enormous quantity of sand and silt brought down from the interior is heaped up or deposited over large areas where the river passes through lower ground. This deposit is of great depth, and maintains a steep face towards the river. It is well exemplified at the bridge, where we sought its narrowest limits. It often forms valleys a mile in width on each side of the river, and these valleys are covered with forest trees and underwood, and a vegetation of the most luxurious description. They are intercepted by deep rocky glens or water courses, where it is dark at noon, and by numerous lagoons and large lakes and marshes teeming with snakes and crocodiles, carpinchos, and other animals. The woods are full of beautiful birds, and are the haunts of pumas, tiger-cats, and other wild animals, with a large number of wild cattle escaped from the open camp. The rich vegetation which covers these extensive alluvial deposits is characteristic of tropical rivers, and is called the Monte. It invariably marks out the distant track of every stream and river, but it is especially prominent in a rocky, sandy desert, where the country everywhere else is totally destitute of vegetation, and it has here assumed vast proportions, on account of the magnitude of the valleys through which the river runs and the excessive height and frequency of the river floods. It is rare that any human being, except occasionally an escaped felon, enters these grand and prolific solitudes. From time to time the cattle from the plains stray into these woods, as they are dotted with luxuriant glades of pasturage, and in such cases they seldom return, and are lost to the owner; but occasionally, in order to recover the cattle, a battue is organized, a numerous troop of mounted gauchos scours the rocks and dells, and in a portion of this Monte near our cottage no less than 2,000 head of cattle were recovered in an expedition of this kind during our residence at Arapey. Such battues are made every four or five years, and to a lover of nature it is well worth a journey to South America to be present at such a chase.

It was among these woods, and rocks, and lakes, and caves, and marshes, crowded with the most gorgeous subtropical botany, and full of marvels in natural history of the deepest interest, that we wandered from morning to night, and often far into the night, to listen to the magic music of the woods. The deep sigh of the alligator, the metallic sound of the frogs, like gangs of stonemasons picking granite; the scream of large owls; the roar of the puma; the splash of the carpincho, as he plunges into the water; the rush past of some straggling ox disturbed by our presence, and breaking its way through the dense thickets, with a host of other mysterious noises, of which we could give no account, afforded intense excitement, often not unmixed with fear, but always of the most enjoyable and impressive description. A gun was an invariable companion, and it was impossible to thread such wilds without a compass, and even with this assistance some large lagoon or deep ravine would often intervene and intercept our way; but the extra fatigue only enhanced the little luxuries of our modest home.

We often at dusk descended the steep and stony bank, through a thicket of thorny talas and acacias, and embarking in our little boat rowed up and down the river bed, among shallow rocky pools, as far as the rapids, which were audible for a long distance, and limited our excursions. We were then embowered between two precipitous heights covered with forest trees, which were smothered in begonias, passion flowers, air plants, and creepers. The river pigs tumbled in from the banks and swam across the stream; flocks of screaming parrots flew by overhead, and pigeons cooed on every side among the trees. This was in the normal quiet state of this clear and gentle river, but in its angry moods in times of floods it becomes a furious torrent, rising thirty or forty feet with great rapidity; it then submerges all these wooded lands, and covering high trees, or wrenching them out of the clayey soil, it hurls them down in tangled rafts among the rocks and underwood, and its noise and turmoil are indescribable. The alligator and carpincho then retire, and seek shelter among the quiet lagoons and inland lakes, which, after becoming more or less parched up by the dry and burning atmosphere, are again filled and often remodelled by the movements of the alluvium bed, and the *débris* brought down by the muddy waters. On these occasions the cottage trembles on the solid

## The Flora. 229

rock, and we look out from the little tower upon a vast and turbulent lake, with the lofty bridge and railway banks slightly raised above its turbid waters.

The botany was in keeping with the dry subtropical character of the climate, the fertile nature of the sands and alluvium, the occasional submergence of low lands by floods, and the liability to destruction by rodents and cattle. Plants that adapt themselves to these peculiar conditions thrive in wanton luxuriance. Within fifty yards of the cottage we could enter the Monte and revel in a flora entirely new and peculiar. Our first visit was in July, or midwinter, but our long sojourn included the summer months of October, November, and December. The soil around the cottage is a dark-red sandstone, covered with very little sand in the hollows, but over large areas quite bare, or covered with large stones. The rocks fall off precipitously near the river, and a rocky ravine brings down a little stream near the cottage. Large patches of this soil are covered with a dense turf-like bed of white and scarlet verbena. The portulaca, with its showy tufts of beautiful flowers, thrives on the hot rock close to the door, interspersed with various lilies, sisyrinchiums, and irises, solanums in endless variety, some like bunches of primroses, others in the form of climbers and shrubs

Œnotheras, salvias, sweet-scented petunias, and especially the lovely macrosyphonia, withstand the treading of the cattle and the intense solar heat, and flourish all over the rocky plains; the other plants that hold their own are grasses and rushes, or plants with wiry, linear, or woody foliage, but none with succulent expanded leaves.

The talas, euphorbias, acacias, and other thorny trees near the house, are tufted with curious mistletoes, and covered with air plants, "pourretia," and "oncidium," and luxuriant creepers, bignonias, asclepiads, passion flowers, tamus, clematis, and a host of others.

But it was more especially in the great Monte, and among the lagoons and marshes, that the botany was so rich and so peculiar.

The showy ayenias, pavonias, and other mallows covered the rocks with lovely flowers. Jussieœ and other onagraceœ surrounded the pools; asclepiads, passion flowers, and bignonias, with gorgeous flowers, overtopped the tallest trees with clouds of bloom. Turneras, white, and pink, and red, with their stained petals, covered the open sands. Gentians floated on the brilliant pools, but the most striking features were the great beds of pontederias, with their large spikes of blue and azure flowers, which covered the deeper pools. The alligator often raised his snout among their large green floating leaves.

The woods are rich in laurels and myrtles, with their perfumed flowers. The similax, with its curious clusters of leaves and tendril-like stipules, clings immovably to the bushes; but the wildest climbers and most showy plants are the papilionaccoe, among which the erythrinas are forest trees dispersed all through the woods and covered with scarlet flowers.

It is very singular to observe how the principle of natural selection has peopled these Montes and parched plains with plants so peculiarly and beautifully adapted to the exceptional characters of the locality. Overrun constantly by rodents and cattle, and subject to long droughts beneath a tropical sun, at other times subject to floods, which sweep away whole islands of vegetation, no ordinary plant could exist. As a protection against cattle, plants, as a rule, are here highly poisonous, or thorny, or resinous, or indestructibly tough. Hence the great development of talas and acacias, euphorbias, solaninums, and laurels. The buttercup is replaced by the bitter gentian, or the oxalis, with its viviparous buds. The passion flowers, asclepiads, bignonias, convolvuluses, and leguminous plants escape the floods and animals by climbing the tallest trees, and towering overhead in floods of bloom.

The ground plants, porlulacas, turneras, and

œnotheras are bitter and tough, and strike their roots into the arid red rocks. They flourish without soil and without moisture, or even dew, and it is singular that these plants immediately wither when gathered and placed in water. The pontederias, the great alismas and plantagos, with grasses and sedges, derive protection from the deep and brilliant pools; and although the Monte may at first sight appear to be a wild scene of confusion and ruin, on closer examination it will be found far more remarkable as a manifestation of harmony and law, and of the marvellous power which plants, like animals, possess, of adapting themselves to the local peculiarities of their habitat.

## CHAPTER XIV.

### *NOTES AT ARAPEY.*

AT the time of our arrival at the cottage in October the country was infested with swarms of locusts, which destroyed the crops and caused great distress. Trains on the Campana travelled with great difficulty on account of the greasiness of the rails, arising from their destruction by the tram; they appeared in dark clouds, like a coming thunderstorm, rapidly devouring every vestige of vegetation, down to the hard and woody pampas grass. They lay so solidly on the ground as to be ankle deep, and when washed up on the beach the putrifying masses poison the atmosphere. They were bred in large numbers in dry sandy spots about Arapey, where the eggs had been deposited, and the only chance of extirpating them appears to be the destruction of these deposits. We often came across a young progeny, covering 120 or 130 square yards of ground, rapidly destroying the grasses and thickly covering every shrub as they moved on. A column on one occasion made straight for the cottage, but by igniting the dry grass we turned them aside, and the poultry assisted in the attack. The Monte near the cottage was the haunt of numerous beautiful birds, especially humming-birds, that darted about among the flowers with extraordinary agility, and afforded us constant amusement. Several mocking-birds, " callandria," built their nests among the boughs of the cattle-sheds, to avoid the iguanas, comadracas, and snakes, which were disagreeably numerous everywhere. These birds became excessively tame and familiar, and amused us with their thrilling song, but annoyed me by a habit that I could not eradicate, of settling on my eagle's feather wind-vane, and swinging round constantly with the wind. Two pair of lovely Widah birds, with white bodies and black borders to their wings, became very familiar pets. Some of the birds we shot were very handsome, and among them a fine eagle and a lovely small white crane, besides some gorgeous woodpeckers and magpies. A good deal of time was occupied in protecting our poultry from the vultures and caranchos, that were always on the look-out; we never left the house with the gun without being followed by a flight of the handsome but noisy "ptero pteros,"who frightened everything away.

Two boys with great pluck shot a fine puma close to the cottage; it had been marauding among the cattle, and was a powerful animal, measuring 6 feet from the nose to the tail, the tail measuring 2' 2": we were glad to be rid of such a neighbour.

As soon as I had settled down in the cottage, which was in a very exposed situation on the hill, I occupied a day in improvising a lightning-conductor. The conductor consisted of some No. 8 telegraph wire which we had in stock, carried on the flagstaff above the top of the tower, but the difficulty was to find an earth on this hard, dry, non-conducting rock. The same difficulty had been experienced at the station two miles away, where a well had been dug which was totally inefficient, and we had been compelled to put up a special wire to find an earth in a marshy spot more than a mile from the station. I was obliged to carry my wire down to a small stream which passed near the cottage, but it was a laborious operation to cut a channel in the rock to avoid disturbance from the cattle, so we covered it with the boulders which lay all around. It was not long before I enjoyed the security it afforded, for on the 26th of October we had a hot north-east wind, with temperature in the shade 96°. The sultry, stifling heat rendered sleep impossible, the minimum of the evening being 84°. The hot gale drove dust and stones against the windows, and shook the tower. This was followed by a lull of singular tranquillity, and then a hurricane from south and south-west, with torrents of hail and rain and thunder and lightning, which lasted far into the night. We counted thirty-six flashes in one minute; the rain, indeed, had not ceased at sunrise. Previous to the storm, huge mountains of cloud rose suddenly in the east, and the line of demarcation between the upper and lower currents was beautifully defined by mottled light wisps of cloud, like the tongues of vapour on the lake. The Arapey rose suddenly, and the next day was devoted to fishing and botanizing under a lovely sky.

The lightning was extremely vivid, and far more copious than in Europe, and though it does not attain the high degree of tension that it would in dryer air, I have been surprised at the few fatal accidents that are recorded; it must be due in a great measure to the sparseness of population and imperfect record, for it is no uncommon thing on fifty miles of railway at Campana to have thick wire fused and four or five telegraph poles destroyed in a single storm. The destruction of cattle is large, but where a mare is only worth about six shillings, and the plains are strewed with carcases, it excites very little notice. A singular accident occurred at Arapey which might have had more serious consequences. At about 120 yards from the cottage, a wooden shed, ten feet square, with a corrugated iron roof, had been used as a gunpowder magazine during the blasting of the rock. It consisted of four upright posts with boarded sides, and contained while we were there one cask of blasting gunpowder unopened, and another cask half full with a board laid over it; the casks stood in the middle on the dry hard rock. During a thunder-storm the lightning struck the roof and descending the four corner posts splintered and demolished them, but probably escaping over the wet surface of the rock, it left the gunpowder uninjured. This very slight shelter would probably have protected an inmate; but I witnessed an example in Wales in which a substantial house was totally wrecked, the roof lifted and displaced, the staircases entirely destroyed, and solid stone landings ploughed up. The bell-wires were all melted, and even a watch left no other trace of its existence than a dark stain on the wall where it had hung. One inmate was killed in bed, and another in the same bed was seriously injured. My little lightning conductor might in that case, by drawing off the electricity from the air, have prevented the lightning from attaining sufficient intensity to strike that particular spot, but would have been absolutely useless as a protection against the current itself when once determined.

In the daily ride to the station our servant caught and brought home a very singular and ferocious pouched animal which he had caught on the railway bank; it was called by the natives camadraca. It was the didelphys Azarae of naturalists, belonging to an order Pedimana, exclusively American. It is an animal very common here, and terribly destructive among our unfortunate poultry. It was a female, with twelve young in her pouch. On turning out these little active new-born savages, instead of any attempt to escape they flew at our feet, snapping at our shoes with their sharp teeth and formidable jaws; we lost all compassion, and with difficulty killed the lot with sticks. The mother neither gave nor asked for quarter, but flew savagely at the bars of her cage and at everything, and we were obliged to despatch her. I never came across an animal so determinately savage, though not larger than a beaver.

Another great enemy of the persecuted poultry was a great lizard, "Podinema teguixin." We killed three of these monstrous saurians, who found shelter under some heaps of timber near the cottage and made fearful inroads upon our eggs and chickens. The largest was a formidable creature, and measured 4 ft. 2 in. in length. The jaws were armed with a threatening array of sharp teeth, and like the alligator it is capable of inflicting a very severe blow with its massive tail. The tongue was highly developed, and the hard rough skin was handsomely marked, and resembled the old shagreen used for spectacle cases; it was remarkably tenacious of life. The tail is said to be good eating. The armadillo dasypusvillosus, which is very plentiful all over the pampas, is a very favourite article of food, and is sold in the markets.

We came across a very interesting animal, by no means uncommon—the cuati or nasua socialis. It belongs to the bear family. It is a powerful creature, but is very easily domesticated, and in that state is very fond of eggs. A friend at Salto had one in her pattio, which had a singular habit, whenever we gave it a piece of soap, of immediately rubbing it on its large tail, and working it with its paws into a thick lather. It was extremely docile, and fond of being petted; it would not allow a dog to come near it, and was destroyed for attacking a large and favourite retriever. Another animal frequently seen on the banks of the Parana, the nutria "myopotamus coypas," is generally found in the same spot as the other large rodent, the carpincho, which we have described as being so common. In the Arapey we could seldom look down the river in the evening without seeing three or four carpinchos crossing the water, or plunging in with a loud splash from the banks. It is a large animal, not unlike a pig without a tail, and the natives call it the sea hog. This very remarkable animal, the "hydrochoerus copybara" of naturalists, is the largest of all rodents; it has large nails, and is webbed, and is said to be the favourite food of the tigers that inhabit the river banks.

I have elsewhere suggested that the absence of trees in the pampas is due to the very large numbers of rodents which inhabit South American plains. In addition to the nutria and the carpincho on the river banks, there are many species of rats and mice, several species of pampas hares, and numberless biscachos, that burrow in every dry hillock all over the camp.

The only creatures that were really troublesome were the snakes; they were so numerous that we seldom had a ramble without killing several, and were always in danger of treading on them; we knew spots where we could always find them. Most of them were, I believe, harmless, but some were of considerable size, and when attacked would attempt a courageous defence; the dogs were continually startled by them, especially near the lagoon. We brought a great many to the cottage. A small and very pretty snake was very common, with three black lines down the body. There were two or three kinds of large green snakes, with rows of sharp short teeth. The very handsome coral snake, beautifully marked with coloured bands, was very plentiful. I dissected the poison fangs from several of these snakes; they lay back enclosed under a membrane in the upper jaw, but were too small to have pierced through a gaiter. They were beautifully adapted for instilling poison into a wound.

There was a channel in the body of the tooth; through two-thirds of its length poison was then distributed by a simple groove extending to the sharp extremity. There is a double row of sharp teeth in the lower jaw, all sloping inward. The head is small and narrow. I have seen this snake spring at the stick with which it was disturbed.

We one day killed a very large and powerful snake; quite distinct from any other we had seen, which I believed was the vibora de la Cruz, or trigonocephalus, which is a very dangerous serpent. We hung it on a tree, intending to take it home on our return, but when we came back at night it was gone, and had doubtless furnished a meal for some of the vultures or hawks who were always on the look-out for such a chance. A stray rattle-snake is occasionally found.

The alluvial banks and plains extended far back into the camp on the opposite side of the bridge, and close alongside the line there is a very large and deep lake, embedded in a forest of trees and evergreens; it was tenanted by water fowl, river hogs, otters, and alligators, and large fish. I often stood on the high viaduct in the evening, watching the pastime of this curious fauna, while large birds swooped over the forest or uttered their peculiar cries among the trees. On one occasion I saw three alligators lying together on the bank close alongside the viaduct; I have often shot at these armour-plated giants within a few paces, but it never appeared to have the slightest effect; they generally plunged lazily under the water among the great beds of pontederia and rushes. They are so inconspicuous, as they lay basking among the underwood, that we sometimes approached very near without seeing them, but their presence was often indicated by the peculiar deep sigh which they utter; their eggs are frequently found in the sand and mud. The largest we saw was about ten feet long, but they are said to be sometimes eighteen feet long. The dull large eye and the formidable jaw and teeth gave them a very ferocious aspect, but, though they will kill the dogs, I have never heard of their attacking a human being.

Although the Monte swarmed with mosquitoes and a terrible stinging gadfly, they never left the valley, and we were unmolested in the cottage; our only enemies were a few scorpions under the stones, and the tarantula spider, of enormous size, which we often killed in our bedroom; there is also a small spider, far too common, whose bite has beyond all doubt proved fatal in several instances. An interesting memoir on these spiders has been written by the professor of natural history at the University at Cordova. But a far more serious pest is the minute becho Colorado "tetranychus," a species of harvest bug, that swarms among the dry grass and bushes, and burying itself in the flesh, principally about the legs, produces the most intense irritation; both in Buenos Ayres and Arapey I have been laid up with the inflammation produced by these minute creatures, and many people during the summer months are altogether unable to live away from the town. The remedy, with those whose flesh will bear such treatment, is to rub the pustules with a brush or an empty ear of maize until they nearly bleed, and then apply Florida water or some other strong spirit; there are other remedies, such as glycerine and carbolic acid, or vinegar, but great care is requisite, to prevent serious sores. The only really effective preventative is to anoint the legs with olive oil, cold cream, or any grease every day before venturing among any vegetation.

I took some pains to ascertain the cause of irritation in the bite of several of the diptera, especially the culex autumnatis, or mosquito; I could detect no poison glands, and I think it is largely due to the form of the wound; the insect inserts, through a very small puncture, two lancets, which are opened after insertion and worked round in the wound; a conical hole is thus produced, in which the capillaries have been destroyed in order to supply the blood which is drawn up by the sucking organs; on withdrawing the lancet the small puncture immediately closes and heals, leaving an inflamed and conical wound, full of extravasated blood. The natural food of the culex would be the juices of the river-side plants, which are thus effectually extracted.

## CHAPTER XV.

### *METEOROLOGY AT ARAPEY.*

IT will be seen that the meteorology of the Arapey is no less remarkable than the singularly tropical character of its fauna and botany. Its most striking features are the totally abnormal heat and dryness of the climate, and the great difference between the temperature of night and day. I have already alluded to the configuration of the continent generally as affecting climate; we are here still surrounded with the same treeless plains, but they consist of a hard, red, brittle sandstone rock, covered with loose *débris*, totally devoid of any visible pasture during the summer, and absorbing the sun's rays with such avidity that during the greatest heat the ground obtains a temperature of 130°, and stones were disagreeably hot to the touch. At this temperature, with a difference of 31° between the wet and dry bulb, it was marvellous to see luxuriant tufts of œnothera and petunia, portulaca and verbena blooming luxuriantly in this dusty desert, and in the evening filling the air with their fragrance. Such an atmosphere was to us almost insupportable; we could only lie still in our hammocks, with doors and windows open all the night. Every article we touched felt hot. The Monte was lit up by streams of fireflies, reminding one of a distant Chinese feast of lanterns; they darted round the cottage and under the verandah, and then, soaring in the air, were indistinguishable from the hazy stars. The house was filled with crickets, and hecatombs of moths and beetles and gnats fell round our lamps and covered the tables. The dried vegetation of the pampas was on fire for many miles all round the horizon, and considerably affected the temperature, and flights of locusts came driving like hail against the house and windows. So excessive was the dryness, that old and well-seasoned theodolite and microscope cases were distorted and split, and the timber hut cracked with sharp reports.

The dogs and poultry seemed to suffer more from the hot breeze than we did, and appeared quite helpless and prostrated as they lay on the dust in the shade, but the horses in full sun bore it without any apparent concern. The evaporation being so excessive, the lagoons and ditches were dried up and the water-plants and grasses annihilated; the alligators wallowed in the hot mud nearer the river banks, but the great creepers that covered the forest trees were luxuriant with legumes and winged seed, and gigantic fruits, and thick foliage, as though anxious to shelter the Monte with an impenetrable canopy of flowers and fruits.

We had to obtain our water from the rocky pools which now constituted the river bed, but by tying a wet towel over our water bottle, and swinging it in the doorway, we obtained water of refreshing coolness. In this hot dry wind I have on several occasions lowered it twenty-five degrees below the temperature of the air. This was the atmosphere in which we passed our Christmas Day at Arapey.

The temperature of the day was 101°, and of the night 85.° In making some experiments on radiation, I discovered a practical means of utilizing its effects which we were glad to adopt. Our high temperature being due to the heated rock beneath, I desired to isolate a thermometer from the ground, and expose it solely to radiation. This I effected by placing it on a feather bed laid in the open air, surrounding it with pillows or any other non-conducting materials; I found that a thermometer at 86° fell rapidly to 68°, which was a greater effect than I had anticipated.

We were not long before we put this discovery into practice, for by placing our beds in the evening outside on trestles, we enjoyed the cooling influence of the space above us, with very sensible diminution of the terrestrial influence.

These hot plains, with their dark rock, exposed to a vertical sun without a vestige of shade, are local furnaces continually pouring their heated columns of parched air into the cooler space above. Close alongside lies the great ocean receiving the same amount of vertical heat; it, however, resolutely refused to be warmed, and pays the penalty by continued loss from evaporation, the heat being converted into vapour. The effect was very evident on the climate, and disturbed those regular alternate periodic inroads of equatorial and polar air which were so marked a feature at Buenos Ayres. The great river system forms a natural boundary, which separates Uruguay both geologically and meteorologically from the plains of the pampas. Uruguay is naturally an extension of the tropical regions of Southern Brazil, which here attain their lowest southern limit. The winds are consequently more variable, and the great storms that retreat westward from the Andes are in a very striking manner diverted or dispersed. They are either thrown northward up the great river valleys, where they find the requisite amount of vapour necessary for their propagation, or they are deflected by the ascending heated columns, which form as effectual a

## The Temperature. 249

bar to their progress north as a range of mountains, and exercise a very singular effect in lifting the polar currents into higher regions. The prevalent winds were from the east, due to the higher temperature of the land, which also prevents condensation, while long droughts consequently are frequent, and dust storms take the place of rain.

As an instance of normal weather, a constant east wind prevailed from the beginning of November till the 16th; it was a strong wind, rising at night to the force of half a gale from the south-east; this increase at night was doubtless due to the reinforcement it received from the sea-breeze that prevails in the evening all along the coast. The general mean temperature from November 2nd to November 16th was, day maximum 76°; night minimum 52°; day temperature of heated rock 102°.

The wind decreased in violence on the 16th, and changed to west, with lightning and heavy rain; this was succeeded by a temperature of 92°, the difference between the wet and dry bulb being 32°. A striking proof of the presence of vapour above was afforded by the fact that the black bulb in the sun was only 108°. The reverberation of the thunder among the clouds during the storm was also very striking. We had a dust storm on the 2nd of December, which exactly resembled the one at Campana already described. The heat was most oppressive in the evening, though the thermometer fell from 93° in the afternoon to 82° at 7 p.m. We observed a heavy cloud in the south-west, with constant summer lightning; we paid no further attention, but sat down to dinner, when suddenly an intense thrill of cold startled us from our seats; we all knew instinctively the cause, and began closing doors and windows as rapidly as we could; the servants barricaded all the back of the house, but before we could make all secure, we were wrapt in deep darkness, and the storm burst over us in all its fury. The noise and turmoil of these storms is indescribable; the cottage visibly swayed before the blow; clouds of dust filled the air, and a tide of chaffy seeds of the flechilla, a grass which covers the pampas, rolled over us like a sea, filling the bedroom, before we could close it, to the depth of two feet. We could see nothing, but the storm fell on the back of the house, from south-west. The lightning was incessant, but no thunder was audible among the uproar; it was followed by a very little rain, and afterwards, at 8.15, by lightning of the most vivid description, which illumined the Monte and the river valley all round with brilliant streams from west to east. The gale swept our remada or cattle-shed bodily away, with upwards of a hundred head of poultry at roost on the top; numbers of them came screaming to the door of the house in the darkness for protection. The absence of rain was singular, but as the temperature of the air before the storm was 93°, and the dew point 54°, while the temperature of the whirlwind was 76°, the formation of rain beyond a few drops was impossible, even if the whirlwind was saturated, as the vapour was almost instantly absorbed in our dry hot air. A marked want of force in the sun's rays preceded the storm, and distant thunder became singularly audible; the change of wind, as the cyclone passed over us, was instantaneous, precisely as at Campana, but rain being impossible, it become purely a dust storm. Its violence again extended over a very large area. Boats were sunk and much damage done at Salto, where passengers in the open air were helplessly hurled away.

This vortex rotated normally around its centre with the hands of a watch, but travelled bodily, directly against the wind, like ordinary storms, of which it was only a specific phase.

These tornadoes often arise on the confines of two currents; they appear to wander over enormous areas, disturbing half the western hemisphere with their capricious motions, though their separate nuclei are small. There are, probably, several of these vortices, simultaneously wandering on, constantly dissolving and reforming, as we see them in a mill race on a river skirting the central rapid torrent. They supply the mechanical force necessary for blending the currents and flashing them into rain; but when they met the dry parched air that prevailed at Arapey they were powerless to produce rain, and expended their energy in raising columns of dust and destroying a few roofs and remadas. As they travel along they hurl together all the layers of atmosphere they meet with, picking up the moist strata that overlie the river valleys, or marshy grounds, or forests, converting all vapour into rain and heat with almost explosive violence; and more especially deluging all such localities. The heat set free by condensation rises like a draught through a tall chimney in the centre of the vortex, surrounded by a cylinder of cold air, which flows down spirally from above to supply the vacuum. Such a system soon dies out on a dry arid plain, the only fuel it obtains being the heated dust. Hence the invariable course of each storm is to follow rivers or wet valleys, where they meet with the requisite reinforcement of moisture. The thrill of cold which always so startles us is the descending current.

The analogy between these tornadoes and the ordinary storm is thus very close. It is in fact simply a torrent and its whirlpools.

## Tornadoes. 253

The barometer fell before the hurricane, as the surrounding air was drawn away into the vortex, from 29·58 to 29·45. It rose to 29·60 when the storm reached us. The descent of the cold column gave the maximum pressure, from which it rapidly recovered. The whole phenomenon is another illustration of the absolute necessity of mechanical force and of the influence of vapour in the production of a storm. The tornado may be regarded as a huge centrifugal pump, drawing up heated air from the earth's surface, and bringing down cold air from higher regions, like ventilating shafts in a mine; its force is maintained by colossal ascending columns of heat, continually reinforced by condensation of the blended atmospheres, while the descending shaft is represented by the cold spiral columns of external air.

These columns move from left to right round the central shaft, but the whole pump stalks slowly and majestically over the pampas, floating on the great currents in which it is generated in a direction contrary to that of the wind. The direction of its motion is determined not only by its naturally seeking an atmosphere loaded with moisture, like an ordinary storm, but also by another circumstance, which accounts for its anomalous progress contrary to the direction of the wind.

Let the isobars A B C, fig. 10, represent the great

natural cyclone of the atmosphere, and the small circle T, the tornado rotating in the direction of the arrows, and in the same direction as the great cyclone, *i.e.*, with the hands of a watch. It is evident the friction against the denser air, where the barometer is at thirty inches, will be greater than against the current at A, where the barometer is 29·7 inches, and that the tornado must consequently

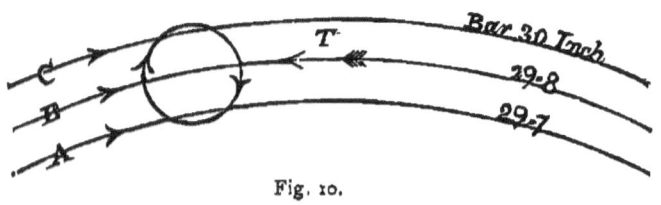

Fig. 10.

roll itself in the direction of the arrow T, that is, to the left, or contrary to the motion of the gentle natural current A B C.

The mechanical violence with which a huge cylinder so constituted must blend together masses of air of all degrees of heat and humidity, furnishes abundant cause for the deluge of rain, hail, lightning and thunder, and every other characteristic of these singular phenomena. It accounts satisfactorily for the fact that such storms soon die out when passing over dry sandy deserts, and attain such terrible violence among the hot saturated vapours of the tropics. The cold weather that follows them is the work done by the pump, which has brought down a whole atmosphere of cold air from above, and raised it to the density due to its lower level, tempering our parched plains with pure and refreshing breezes.

The great feature in all these storms is their self-sustaining power; like trains of gunpowder, when once ignited they run along exploding the mixture they meet in their progress. A very small disturbance may thus originate a very severe storm.

The destructive power of these tornadoes is incomprehensible on any theory of their sudden creation; but all difficulty ceases when we regard them as the accumulated effect of long-continued energy.

I have alluded to the effect of mountain ranges as one source of our storms. The local ascent of heated columns of air, which forcibly overcomes the tendency of strata to maintain their stratification, and diverts terrestrial currents into the upper regions, is another frequent origin of disturbance; the tornado, or the vortices formed along the margin of great currents, is a third, and the further development of either of them depends wholly on the medium in which they originate and progress. The essential cause of them all is the sun's heat, absorbed in evaporation over the ocean where the cloud is formed, but converted into mechanical energy over the heated land when the vapour is condensed, and precipitated in storms and refreshing rains.

A very singular result of this tornado was the deposit of enormous deep beds of the light seeds of the flechilla grass in every dell and sheltered spot. They filled our bedrooms in an instant to the depth of two feet. This terrible plant is the scourge of the pampas; the barbed seed, with its horny processes, and its long arms studded with short hairs, all set one way, possesses a power of penetration that is irresistible. They worked their way through our clothes and gaiters, but they are far more seriously damaging to the poor sheep. They ruin the wool, and, working their way through it, frequently kill the animals that are not shorn by penetrating the flesh, and they are frequently found in the muscle of the joints brought to table. The valley was filled with these seeds; a bank several feet high lined the river for miles; deep ravines and ditches were filled up, and in sheltered spots, among the trees in the Monte, enormous deposits had accumulated, covering the underwood. We amused ourselves for several weeks after the storm by dropping lucifers among these heaps. The flame reached a great height, and the fire spread far and wide, illuminating the woods during the night with sudden bursts of flame, as fresh deposits were reached by the running fire. The flame from a collection beneath the bridge almost reached the high timber platform above. They indicated the enormous area of the pampas, which had been swept by the storm, which still carried them on for hundreds of miles further into Brazil. The distribution of seeds in this manner is an important botanical phenomenon.

## CLOUDS.

The high position of the cottage gave a commanding view of the level camp for many miles around, and its bare surface, with the dried silvery remnants of the spring vegetation, afforded an excellent screen, on which the shadows of passing clouds were vividly depicted. Many pleasant hours were spent in watching these fields of shadow as they drifted past over the pampas, and in estimating their magnitude and speed. It is here that I became convinced of the close analogy between the cirrus and other similar forms of cloud generated on the clearly defined confines of overlying strata of different constitution, and the tongues of vapour I had observed during our quarantine that formed and drifted away and re-dissolved on the surface of the great river. The symmetric forms and changing aspect of other clouds as they passed over could only be satisfactorily explained on the very probable assumption that it is on the outer surface only of cold masses of air that cloud is formed, the centre remaining transparent. Such clouds are therefore shells of vapour, and generally follow inroads of polar air, whereas the dense cumulus and the storm clouds produced by equatorial air is always opaque and solid throughout, however subdivided the mass may become. But the most characteristic and frequent form is the stratified cloud that forms in the horizon. I know of no more certain prognostic than those long, horizontal, clearly-defined, straight lines into which dense clouds range themselves previous to rain.

This appearance is satisfactorily accounted for by the impermeability of the lower terrestrial stratum of air which rests on the earth, and is subject to its direct influence.

The appearance observed is due to perspective. The cloud sailing on this stratum like oil on water preserves constantly its horizontal base. When seen in the zenith, from directly beneath it, the level base is imperceptible, and we see only the cloudy margin as at A, fig. 11.

When the same cloud reaches the horizon, its base on the terrestrial stratum being now brought into view, it assumes the form of a horizontal layer:

The dip of the horizon is so slight that the surface of the terrestrial layer is practically a horizontal line, and the cloud is seen by an observer at O, as in fig. 11, in which it will be seen the base subtends a less and less angle the farther it is re-

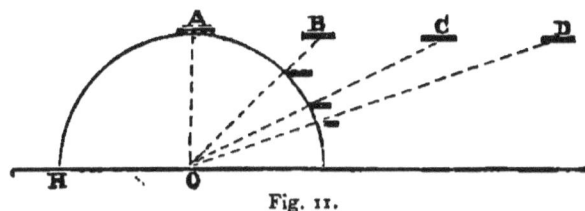

Fig. 11.

moved from A, and is ultimately brought into view as a straight line.

The observer refers the various positions to the visible hemisphere above him, and really sees the clouds as though on the circle A H, where they assume the form of straight lines on the horizon.

These appearances, which are almost of daily occurrence, on a small scale, afford occular demonstration of the defined character of the line of junction of lower and upper strata.

It is by no means uncommon in stormy weather, when the line of separation is low, to see the clouds on the upper stratum moving in a direction directly opposite to those below. At greater altitudes, as at Teneriffe, this is often a permanent phenomenon.

This boundary line between two currents does not necessarily produce clouds, as both are frequently far removed from saturation, but it often causes that peculiar haze which arrests the force of the sun's rays, and in the night obscures the lesser stars.

The height of clouds depends on the vertical depth of the terrestrial stratum, which again depends largely upon the temperature of the earth itself. On this account clouds are, as a rule, higher in summer than in winter, but gales of wind, by disturbing the terrestrial stratum, often lower their height, and are succeeded by rain, when the terrestrial stratum is totally swept away, and they come into mechanical contact with the earth itself. A single stratum of cloud is generally of very moderate depth, varying from a few yards to three or four hundred. When condensation takes place, and rain commences, the effect is accumulative, as we have seen in storms, but even with a gentle rain, cold air is brought down from higher regions, and a more rapid condensation is engendered, and more cloud is formed.

The above and other deductions are taken from my Arapey notes, but they do not apply to every form of cloud. These results would indicate three very distinct classes—1st, the stormy cloud, formed by volumes or masses of heated and saturated visible vapour floating through a transparent medium; 2ndly, the light shell or superficial cloud formed on the external surface of cold masses of air floating through a transparent medium, and, 3rdly,

## Effect of Shadows. 261

the horizontal stratified cloud formed at the limiting boundaries of two adjoining transparent strata. The true storm-cloud arises from the forcible blending of either or all of these forms, with a general condensation of the vapour itself, produced by the mechanical energy of the storm at the earth's surface. The true nimbus is therefore a mere outward and visible sign of the mighty transformations that are in progress.

The lifting and falling of clouds is not a bodily motion of the whole mass, but is more generally caused by absorption on one surface, and new formation on the other; this was evident at Rio, where a cloud remained stationary on the summit of the Sugar Loaf during a gale of wind. I will only add that the thermic features of passing clouds may be recognised very accurately by a thermometer screened from the effect of the earth's radiation, while every one must have bodily experienced the frequent change of temperature due to a passing cloud.

On the 24th of December the thermometer was 103°. The parched hot wind was like a blast furnace; the nights were cold. These extraordinary changes, which are of constant occurrence during the continuance of the same east wind, afford striking evidence of the direct effect of the sun's heat. The temperature of the rock was 130°. The obliquity of the sun's rays is of course a principal cause of the cold of our northern climate, as compared with the tropics, the same angular amount of solar radiation being distributed over a larger area as we recede from the equator; but it appears to me there is another cause of the great difference of temperature, of very considerable importance.

At Arapey, the sun at Christmas—equivalent to our midsummer—at noon is about $7\frac{1}{2}$ degrees from the zenith. The shadow of a tree or tower 100 feet high will therefore be the tangent of $7\frac{1}{2}°$ or 12·1 feet in length, whereas the shadow of the same tower in London would at noon at midsummer be more than half the height, or fifty-three feet long, and in midwinter it would at noon be no less than 373 feet long. This enormous difference in the length of the shadow from every projection must have great influence on the temperature acquired by the solid earth. This difference must be greatly increased at Arapey, where we have a vast plain without any trees or vegetation, whereas enormous tracts in England are in constant shade.

The tree no doubt lessens the nocturnal radiation of the soil, but only in proportion to the actual space it covers, whereas its shadow sweeps over large areas all around. Such a difference cannot fail to exercise an important effect on the temperature of the terrestrial stratum. The absence of

## Influence of Shade. 263

humidity is all also greatly due to the want of shade and vegetation. We had no dew, and the difference between our dry and wet bulb thermometer reached 31°!

The mean day temperature of this terrible Christmas week was 90°3′, and of the night 58°: maximum 103°, minimum 45°.

This excessive heat continued till our departure on the 31st, in another hot dust storm without a drop of rain; but parched and barren as the camp had now become, we left our peaceful cottage with sincere regret. The mighty Uruguay was now a shallow rocky stream, and its navigation was difficult even with the little steamer that took us back to Paysandu.

## CHAPTER XVI.

*PARAGUAY AND ITS HISTORY.*

IN the very centre of the South American continent, on the tropic of Capricorn, and 1,100 miles up the Parana and Paraguay, there is a little Mesopotamia, singularly distinct from all the surrounding provinces, and equally remarkable for its unrivalled fertility, its picturesque beauty, its magnificent climate, its gentle, brave, and docile population, and its brief and cruel history. The annals of Paraguay, though extending over so short a period, furnish a condensed epitome of human history of the most varied and eventful character. They are records of the highest virtues and the blackest crimes of which man is capable. The Christian virtues of docility, endurance, and submissive resignation have never been more prominently exhibited than among the descendants of the gentle pagan tribes who inhabited these fertile and lovely provinces, while, on the other hand, the dominion of their Christian conquerors has furnished a page of human history unsurpassed in its record of wanton atrocity and cruelty by any pagan annals. The ultimate destruction of this rich and thriving population by a sensualist, savage despot is an historical episode of the darkest dye and of the deepest interest. Ever since the time of Pizarro the whole history of South America is little more than a record of Spanish misrule, plunder, and cruelty, inflicted on an inoffensive and defenceless population ; but in this last, and it is hoped closing scene, the usual incentive of plunder was wanting, and with no other object than the gratification of wanton barbarity and ambition, this brave and faithful race was so completely annihilated that even at the present moment the remnant of the population, reduced in four years from upwards of 1,200,000 to 300,000, numbers eight females to one male.

The original inhabitants were the Guarani Indians, the remnants of an agricultural people, pre-eminently superior to the surrounding savage tribes, and remarkable for their gentle character, their developed and wide-spread language, customs, and religion, and their greater progress in civilization, as evidenced by the remaining monuments of their prosperity, which attracted the early Spanish adventurers.

The city of Asuncion, on the banks of the Paraguay, was founded by the Spaniards in 1535. The country was soon overrun, and was ultimately peopled by a mixed race, mainly Indian or negro, but partly Spanish and partly mulatto, from the marriage of Spaniards with the original inhabitants, for the descendants of those marriages pertinaciously admit no more Indian blood. At a later date the extraordinary fertility of the soil, and the character of the inhabitants, attracted the Jesuit fathers, and as the country fortunately contained no precious metals, these missionaries were endowed by the Spanish government with unlimited independent authority. Among so kind and docile a people, they found a field for their labours totally unexampled in religious history.

Entirely incapable of independent action, the Indians submitted like children to the gentle rule of these intelligent and highly-educated instructors, who established, in the beginning of the eighteenth century, a purely sacerdotal government, as thorough and as uncompromising as that of the Israelites of old. Instead of being employed as slaves, the Indians were educated and instructed in the useful arts and trades. Public amusements of every description were provided for them; justice, order, and law were strictly maintained. Ecclesiastical edifices and mercantile establishments of the most substantial and imposing character were erected, and the country was cultivated with skill, and enriched with villages, public roads, and gardens, and extensive plantations. The oral language of the Guarani tribes was converted into a written language, and taught grammatically in public schools, by printed grammars, dictionaries, and missals, in the new language. Under such tuition a large population was brought to a state of the highest military discipline and obedience. But all this material happiness and prosperity was bought at the sacrifice of every vestige of independence. Their new lessons had permitted no development of self-reliance, but had only confirmed that habit of implicit obedience which was the characteristic evil of their original savage life. The result, however, sad, was only natural; so simple-minded, helpless a flock was but too ready a prey for the first reckless adventurer who required such welcome tools.

At the first summons after the recall of the Jesuits, without a thought or a murmur they abandoned their peaceful homes, and did not hesitate to yield the same childlike obedience and fidelity to the tyrants who persecuted and ultimately destroyed them, as to the amiable though narrow-minded priests to whom all their happiness was due. The Jesuits were suddenly recalled from South America in 1767, and Paraguay again fell into the hands of Spanish governors, and lost two-thirds of its population in thirty years. After the revolt of the Spanish colonies in 1814 it became independent from sheer inaccessibility, when a Dr. Francia seized on the dictatorship, and instituted a system of tyranny and espionage of the most oppressive character. In imitation of the merchant Jesuits, he closed the country from all access by land or water, and became sole importer and exporter of its produce. He detained forcibly all foreigners who set foot in the country, and discouraged marriage and religion, and ruled with cruel and capricious tyranny until he died in 1840.

He was succeeded by another adventurer, the first Lopez, who, though he permitted his family to enrich themselves by unscrupulous plunder of this gentle and enduring people, and assumed the most despotic authority, committed but few atrocities, and was averse to the judicial murders which had decimated all the best families under the rule of his blood-thirsty and cruel predecessor.

Though under jealous restriction, he opened the country to foreign trade, and the people, now left alone to enjoy the riches of these fertile valleys again, rapidly attained a material happiness, probably unparalleled in any part of the world. The population was rapidly increased. They were excessively hospitable, and a few days' labour produced all the tobacco, maize, and mandioca they could consume, while they lived in groves of orange trees and bananas, surrounded with excellent pasturage for their cows and cattle. The principal revenue of the country was derived from the yerba forests. The yerba is a tea of singularly refreshing properties, largely consumed in South America, produced from the leaves of a plant of the holly tribe, the Ilex Paraguayensis, which naturally covers extensive areas in Paraguay, and might, under proper regulation, produce enormous revenues.

Lopez died in 1862, and was succeeded by his son, the late Lopez II., the most senseless, cruel, and blood-thirsty tyrant among all the terrible list that have desolated the South American continent. This unnatural despot was bayonetted in a ditch by the Brazilians in 1869, but in the short period of seven years, this prosperous and happy country was so ruthlessly destroyed and annihilated that the population was reduced to 200,000 souls, principally women and children, driven into the woods and marshes. His insensate war with Brazil and Buenos Ayres, his fiendish cruelty, his vanity, cowardice, and rapacity are almost inconsistent with sanity, and would have been impossible with any other population than the brave and faithful but helpless and enduring victims which such exceptional development had prepared for him. His own education appears to have been specially destined to foster his naturally depraved disposition. Brought up in ignorance and profligacy and encouraged in extortion, he was sent to Europe when a very young man by his father, the elder Lopez, with a retinue of dissolute companions, and a revenue almost fabulous in amount. He led a scandalous and disreputable life in London and Paris, where he met with Mrs. Lynch, whom he carried away with him to Paraguay, and to whom he remained permanently attached for the rest of his life. His visit happened during the Crimean war, and it is said he spent some time in the camp of the allies in the Crimea, where he acquired a taste for military power, and resolved on his return to convert his own happy pastoral province into a huge camp. His originally savage and lawless disposition was thus supplemented by all the lowest vices of European civilization. After his death the wretched remnant of this persecuted people, consisting almost entirely of women and children, returned to their desolate homesteads, and resumed their simple agricultural labours. The city of Asuncion was in ruins, and the villages deserted. The revenues of the country did not yield sufficient to defray the cost of maintaining even a semblance of public order. But adventurers of the old type, who had fled during the war, soon returned, and the privilege of plundering this scanty remnant of the wreck led to fresh plots and revolutions so revolting to common humanity, that it resulted in a temporary protectorate by a Brazilian garrison, which imposed a heavy burden on the population, hitherto free from debt. The railway which Lopez had constructed, principally for military purposes, runs through so rich and productive a district that in quiet times the traffic is considerable, but as the country was entirely depopulated, it was allowed to fall into complete decay. The large and handsome stations were roofless, and the bridges unsafe, the permanent way was completely neglected, the extensive equipment of rolling stock was worked as long as it could run without repairs, until at last a single engine and two or three ruined carriages made an occasional run over the forty-six miles to Paraguari, which was often reached with difficulty in a day. In the absence of coin, small tokens used by a local tramway company were used for barter, and public credit was entirely destroyed. In this state of affairs, when plunder at home was no longer possible, as even Government *employés* were unpaid, this unhappy people heard with absolute incredulity that a loan for £1,000,000 sterling had been negotiated in their name in London, followed immediately by a second loan of £2,000,000 in 1872, and that they were saddled in perpetuity with a debt that only ceased to be formidable on account of its magnitude. The professed object of the loan was the extension of the railway and other public works, but no attempt was made even to maintain the slight remnant of traffic on the existing line, which ceased altogether. It is of course impossible to obtain any authentic account of the proceeds of this loan. It is generally stated in Paraguay that the whole amount which fell to the share of the negotiators there did not exceed £30,000. With every phase of tranquillity the country speedily recovers itself with marvellous celerity, and if left alone in the enjoyment of any security, would, doubtless, from its great natural wealth and resources, and the character of its inhabitants, rapidly attain its original prosperity; but the slightest symptom of improvement attracts rival adventurers, with the usual plots, revolutions, and assassinations. It is only four years ago that, the growth of tobacco being found to be highly remunerative, a considerable crop was grown, but as soon as it was ready for gathering, a special law was passed, under the late President, by which it was enacted that the Government should be the sole purchasers, at a fixed price considerably below its value, and to add to this spoliation, it was paid for in an utterly worthless paper currency, issued for the purpose. The President was shot last year in the public streets in the middle of the day by half-a-dozen assassins, not, as might be imagined, from among those who had been so mercilessly robbed, for they bore their loss with their usual resignation, and merely abandoned the growth of tobacco. He was murdered by rival political adventurers, who made an unsuccessful attempt to usurp his envied privileges; some of them were arrested and thrown into prison, where they remained a long time without trial, and to complete this miserable history, they in their turn were killed in prison only a few months ago, on the ground that they attempted to escape. A project has lately been entertained for the admission of the province into the Argentine confederation. The only result must be that these persecuted people will be saddled with a still more permanent incubus, in the form of the usual highly-paid President, senators, and congress men, with ministers of every grade and their host of nominees, while administration in the sense in which we understand it will be totally neglected. A few experienced town surveyors, with a small body of properly organized police to maintain public order and cleanliness, and independent magistrates for the impartial administration of justice, is all that is required; but it is unfortunate that these first requisites of civilization seldom obtain attention among politicians who rely so largely on their eloquence, and prefer loftier and more profitable themes for its display.

## CHAPTER XVII.

### *THE PARANA AND ASUNCION.*

IT is impossible to convey any adequate idea of the magnitude of the enormous rivers which empty themselves into the Plate by any comparison with our own rivers. The Parana is the second largest river in the world. For a distance of 600 miles the two banks are rarely simultaneously visible. The steamer plods her way, day after day, against the current, among endless groups of large islands or through extensive lake-like vistas, sometimes intercepted by shallows, which are avoided by crossing or re-crossing from island to island. The baranca, or cliff, which defines the limits of the old river bed, is occasionally seen, and often forms a prominent object many miles inland; but the actual banks undergo perpetual transformation, as the river, in its continual change of *regime*, ploughs away the alluvium where the stream is strongest, and deposits it in the form of new islands, or rush-grown flats, where eddies or quiet waters allow of its subsidence. As the lands that are so denuded are usually covered with timber, the banks are lined with rafts of timber, or with falling trees whose roots are bared by the current. Cranes, storks, and other large birds sit motionless on these piles of ruin, or fly lazily away as the steamer approaches them; while the carpinchos and other animals flounder about beneath them, and higher up the river the alligators lay basking on every sloping beach.

In going to Asuncion we leave the Parana at Corrientes 686 miles from the Plate. Even at this point it drains a mountainous region exceeding 500,000 square miles in area, but its sources are nearly as mysterious as those of the Nile. All navigation is stopped at the great falls of Guaiva, 1136 miles from the Plate. Azara, who visited these falls in 1788, describes the river as 14,000 feet wide, suddenly reduced at the cataract to 200 feet. Thus confined, its waters break with indescribable fury; they drop fifty-six feet over a plane of granite, at an inclination of about fifty degrees; the clouds of spray form columns of vapour, lit up with rainbows, which are visible for many leagues. A continuous rain is produced by the condensation, and the roar is heard at twenty miles distance. The

## The Parana.

district is inhabited by a low race of Indians. Beyond this point all is unexplored; the Rio Pardo is a tributary at 1,400 miles, and the Rio Grande at 1,820 miles from the Plate, and at this point it first takes the name of the Parana. For technical details of this extraordinary river, see "Hydraulics of Great Rivers," by J. J. Revy, 1874.

The tributaries of the Parana are gigantic rivers. The Salado, the Vermejo, the Paraguay, and Pelcomayo are all nearly a thousand miles in length, and run through vast districts occupied by Indian tribes, totally unexplored.

The field which these prolific regions opens for the ambitious traveller are unbounded, and as the steamers which navigate the main rivers are fitted up with great comfort, and even luxuriance, and as the political conflicts which have so long rendered them inaccessible are now at an end, there is no longer any difficulty in exploring a country which is unrivalled for its beauty and fertility, and which is almost entirely unknown. The journey from Buenos Ayres to Asuncion is performed by the steamers in about seven days. The latitude of Buenos Ayres is 34·36, and of Asuncion 25·20, but the longitude is nearly the same, so that the journey is due north, and is a continuous approach to the tropics.

The distance by the river is approximately as follows:—

|  | Leagues. |
|---|---|
| Buenos Ayres to Rosario | 75 |
| Rosario to Parana | 40 |
| Parana to La Paz | 40 |
| La Paz to Esquina | 22 |
| Esquina to Goyaz | 30 |
| Goyaz to Bella Vista | 20 |
| Bella Vista to Corrientes | 30 |
| Corrientes to Asuncion on the Paraguay | 100 |

357, or 1071 miles.

In the journey to Asuncion, the Parana is ascended as far as Corrientes; the remaining 300 miles is on its noble tributary, the Paraguay.

The increasing temperature of the river water is a fair index of the change of climate. It increases at the rate of a little more than 1° Fah. for every degree of latitude, the temperature of the water in August being 62° at Rosario, and 70° at the entrance of the Paraguay. The waters of the Paraguay are much warmer, being 79° off Asuncion.

The height of the river above the sea is 250 feet at Asuncion, 200 feet at Corrientes, 100 feet at La Paz, and 60 feet at Rosario. The fall is, therefore, over 2½ inches per mile, for the whole distance, or 2 inches per mile on the Parana. Its tributary the

Salado, in the Gran Chaco, falls with perfect uniformity five inches per mile for 700 miles, while the country itself, in a straight line, falls twelve inches per mile. The uniformly increasing gradient with which the pampas everywhere rises, from the river valley up to the Andes, has been already alluded to. Cordova and Tumman are 1,300 feet and 1,200 feet above the sea, while Mendoza and San Juan are 2,400 and 2,200 feet. The scale over which this simple and uniform *regime* is maintained is the most striking characteristic. This is prominently forced upon us as we ascend this mighty river; when travelling night and day in a rapid steamer for three or four successive days, no evident diminution in its width or volume can be detected. We appear to sail from lake to lake amidst a labyrinth of luxurious islands, occasionally threading some narrow channel, to enter again some magnificent reach, shut in by dense and forest-like vegetation, which we soon discover to be only another branch of the great network of streams and rivers and lakes of which the great system is composed, but the actual banks are rarely, if ever, simultaneously visible. The baranca, or old river boundary, occasionally towers up in the distance in the form of tosca cliffs; among the trees on the Entre Rios shore, and at Parana, cliffs of tertiary limestone rise out of the river. They contain beds of shells, and large deposits of a gigantic fossil oyster, the Ostrea Patagonica. This shell conglomerate is largely burnt for lime. The town is built on the high ground, communicating by means of a tramway with the iron jetty on the river. This little geological feature is quite exceptional, the river, as a rule, maintaining its shifting channels through vast deposits of alluvium almost from the tropics to the Plate. The river here is washing away its eastern margin. At Rosario its bed was travelling west. Very few vessels passed us, and though a few large birds were perched on fallen trees, or among the rush-grown swamps, the total absence of any sign of human life gave an imposing stillness and vastness to the scene; but the gradual approach to the tropics, which slightly changed the vegetation from day to day, effectually prevented monotony. A few palms appeared at Santa Fe. The yucca grew on the limestone cliffs, and numerous brilliant creepers towered among the trees, which, with the foliage, gave an English park-like aspect to some of the glades. After leaving La Paz, we skirt the great Indian territory of the Gran Chaco on the left-hand bank for a distance of 600 miles. This prodigious territory, covered with forests and swamps, is inhabited only by wandering Indian tribes. We occasionally saw a group with their canoes on the river bank drying skins, which they exchange, by crossing to the Entre Rios coast, for spirits and gunpowder and tobacco. They also bring bamboos, and grasses, and a few mats, but nothing of any value, or indicative of the slightest skill, or intellect, or industry; no lower type of humanity can be imagined. The lovely climate kept us late on deck, and we enjoyed the imposing magnificence of the southern nights. The whole scene was one of the deepest interest; the only sound was the ripple of the little steamer ploughing her way among these vast and mysterious waters, lit up by a brilliant moon, which shone all round on boundless regions of water and land, totally unexplored, uninhabited, and unknown. The Gran Chaco is a mere hunting-ground for a few tribes of wandering Indians; but even amidst this ocean-like extent of land and virgin forests, they retain one of the instincts of their more civilized brethren, and carry on relentless wars for the sake of territory. I have sat on deck half through the night, catching an occasional glimpse of their camp fires, vainly endeavouring to understand the mysterious dispensation, which, after ages of misery and suffering, leaves no possible future for this helpless and degraded race than that total extermination under which they are fast disappearing; for even when compelled by force to come into contact with civilization, they fall without hesitation a prey to its lowest vices.

It is a singular fact that these races in their natural state should make no progress of any kind, but continue satisfied, generation after generation, with the same precarious means of existence, the same rude bows, and arrows, and spears. The great characteristic of civilization is its "Excelsior," and progress only arises from this constant determination not to be satisfied, but to rise higher and higher. The same principle holds good with education itself; I have often been struck with the fact that a gipsy fiddler plays the fiddle eight or ten hours a day, for sixty years, without playing in the slightest degree better when he dies than when he first began, while another pupil becomes a Paganini in a few years. How many girls at school obtain a certain knowledge of music, on which they make no improvement for the rest of their life! The true banner of civilization and improvement is "Excelsior." We make no more progress than the old fiddler, or the soldier who practises the goose step, by simply repeating that which we can already do; we must, in order to advance, be striving to attain that which is apparently beyond our reach, and it is astonishing, when thus assailed, how little there is that is altogether unapproachable. I have already elsewhere compared civilization to water kept at a high level by the constant attempt to pump it higher, and it is a useful maxim that, how-

## Change of Scenery. 283

ever industrious a man may be in doing that which is a mere matter of routine, his day has been absolutely lost, as regards progress, unless he has devoted some portion of it to the more ambitious attempt of doing something that he could not do before.

The actual speed of our vessel was ten miles per hour; the stream against us averaged three; our progress was therefore seven miles per hour, and the consumption of coal for the whole journey was 150 tons, at £3 per ton. The long column of black smoke, that disfigured the brilliant sky, indicated the extravagant waste of our consumption. Although it was only after entering the Paraguay that the great change of climate became so prominent, yet our daily approach to the tropics was strikingly illustrated, especially on approaching Corrientes; large birds, with brilliant plumage, and a few monkeys in the islands, excited our interest, and on a projecting spit of land we saw two large alligators, slowly gliding into the water. The trees, moreover, had changed their character and assumed darker hues; they were now intermixed with huge clusters of bamboo, towering elegantly like ferns among the underwood. The creepers were innumerable, and covered vast masses of vegetation with their showy flowers. An incident occurred which might have had serious results.

The pilot amused himself when the vessel neared the islands by firing at alligators and monkeys, in the trees, with a Remington rifle, and while surrounded by the passengers, the rifle burst in his hands, throwing large pieces of the barrel and stock all around; fortunately no one was struck, but it was a narrow escape. It was not till we were close to Corrientes that we came across fresh symptoms of civilization in the shape of a dark gaucho, driving some cattle along the river banks. The passengers were enjoying their matté as usual on deck, for the temperature was 73°, and it was late on Sunday evening when we arrived at Corrientes, and strolled through the town. A few Indian women, of hideous aspect, were selling their useless wares, and a few natives were galloping about on horseback; but we were objects of curiosity to a most disagreeable extent, for with very few exceptions, my wife and her friend were the only European ladies who had ventured to visit Paraguay since the war.

It is impossible to imagine a more barbarous medley of huts and squares, and sandy tracks, than those which form the streets of the capital of this large province. The houses are roofed with bamboo, tied into bamboo rafters; the population is 10,000. Each house has a little bit of pavement in front, but no two are the same level or in the same line, the difference of level often being two or three feet, so that it is impossible to keep on the paved path, and the road is a deep, loose heap of hot fine sand, in which the shoe is entirely buried at every step. As the sand abounds with jiggers, which breed in the feet, a walk is the most undesirable recreation one can select, while the dark, wild-looking inhabitants stared with a bold vulgar curiosity that made us begin to doubt the propriety of a much longer extension of our journey, and we gladly got back to the little comfortable steamer.

We entered the Paraguay two hours after leaving Corrientes, and the change of climate was extremely abrupt and striking. The Gran Chaco lay on our left, with its confused and entangled vegetation and its great morasses; but the Paraguayan shore is formed of a sandy beach, wasted down from the sandstone banks; on every spit of sand the crocodiles lay like logs of wood, often two or three together, crawling lazily into the water when shot at with the rifle, apparently unhurt. The creepers on the underwood formed bowers of dense shade, sometimes resembling cut hedges and hop gardens, while trees like large oaks seemed embowered in banks of laurels. The foliage, though generally dark in colour, furnished broad effects of light and shade, relieved by the delicate green fronds of the lofty bamboos and the native willow Salix Hamboldii. The trees were tenanted by monkeys, herons, parrots, and cranes; and all these lovely scenes were more accessible, for, though a mighty stream, the Paraguay has none of the vastness of the Parana, its width varying from a quarter to three-quarters of a mile. We came across remnants of the vast defensive works executed by Lopez, and some of the old telegraph posts rose out of the jungle covered with asclepias and convolvulus; but tigers, pumas, snakes, and other animals abound among these island forests, and sometimes boldly ford the stream.

We arrived at Asuncion August 21st; the shade temperature was 82°, and we were glad to avail ourselves of the tramway which runs from the quay, and avoid the walk through the deep hot beds of sand, of which the roads consist.

The situation of the city is extremely beautiful. Immediately above the town the river is divided into two branches by a large island; the channel on the side of the town widens out into a vast lake of brilliant shallow water, with numberless creeks and pools and wooded islets, resembling the finest scenery in our lake districts, and literally swarming with crocodiles and magnificent birds, and so completely covered with water-plants as to resemble fields of luxuriant vegetation. Among the pontederias and other water lilies the Victoria Regia, with its enormous floating leaves and flowers, grows luxuriantly in some of the shallows, and shelters the numberless tenants that swarm in the prolific waters beneath. The flowers are eagerly sought by the natives on account of their extraordinary beauty, and the plant is becoming scarcer in the immediate neighbourhood of the town. The river bank is a steep wooded cliff of sandstone, and at spots, where streams of water flow down from the higher ground above, the natives have extemporized a rude kind of bath by means of wooden conduits. A crowd of bathers avail themselves of this luxury every evening, and enliven this charming scene with their picturesque deshabille.

The town occupies a bend of the river overlooking this lake-like expanse, with the wilderness of the Gran Chaco in all its sullen, monotonous grandeur on the opposite bank. A journey across the water thus brings us in a few minutes into another world— a wilderness uninhabited and unknown. We landed for an hour, and made our way some distance among the dense underwood and vegetation, but the heat, and the mosquitoes and other insects compelled us to beat a retreat. There is, however, a hut or two, occupied by Portuguese settlers, but it is difficult to imagine how they can hold their own in such a wilderness, or why they prefer this parched and sun-burnt desert to the lovely paradise of waters and fruits and shade that is stretched before them on the opposite shores.

The city of Asuncion is, of course, a wreck, but retains many emblems of its former prosperity. The streets, as in all South American towns, are all at right angles to each other, and are laid out with more regularity than usual. Among other amusements of the tyrant Francia, he unfortunately possessed a small theodolite, which was as fatal as a battery to anything which came within its range. Houses and streets were demolished in the most arbitrary manner, without hesitation or compensation, whenever he had a surveying fit; but as this was done without judgment, natural water-courses were intercepted, proper drainage was impossible, and during tropical rains whole squares were often flooded or submerged. His own palace was in the upper part of the town, and consisted of a square, surrounded by solidly-built, plain, low, single-storied buildings; whenever he entered the town the streets were cleared, and the inhabitants were ordered into their houses by sound of trumpets, and anyone seen at the windows was immediately imprisoned. But so rapidly did the wealth of the country increase, even under the precarious severity of this crushing *regime*, that only fifty years later we find Lopez II. drilling an army of 64,000 well-equipped troops, casting his own cannon, building iron vessels, and erecting the magnificent arsenal which lines the river bank, and is still filled with the

## The New Palace. 289

ruins of modern machinery of the most costly and perfect character. Spacious barracks were erected in the city, an opera house was commenced and nearly completed on a scale of magnificence and magnitude rivalling our largest European houses, and in lieu of the old barracks occupied by Francia, a magnificent palace was constructed on an eminence commanding lovely views of the river, with its islands and lake-like expanse. The ruins of this splendid building form an imposing object in the town, and great skill was shown in its construction. The base is built entirely of large, well-worked blocks of the fine sandstone of the district. The upper building, with its handsome, lofty tower, and colonnaded portico and external galleries, is in excellent stuccoed brickwork; the design is Italian, and exceedingly chaste; the timber is wholly quebracho, a wood as hard and durable as iron, and nearly as heavy. The columns consist of a central nucleus of quebracho, carefully worked, and covered with tiles and stucco. The external balconies and galleries are paved with large stones laid on hard timber joists; the building is now unroofed, the timber has been stolen for use in the town, shot and shell have fallen through the galleries and floors, and a portion of the elegant campanile is carried away; this has been done in wanton mischief, and not with any military object, and is one of the cruel legacies left by the Brazilians and their allies. It is estimated that the building could be restored for about £60,000. In addition to these works, forty-five miles of a costly, well-constructed, and perfectly-equipped railway were completed, from the city to Paraguari, and was in course of prolongation to Villa Rica, a distance of ninety miles; and the remains of an extremely handsome colonnaded station, of the most substantial character, now forms the city terminus. The railway was continued down to the river for military purposes, where massive quays and warehouses line the river bank. When we consider that, in addition to all these works, he carried on a long and costly war, which almost exhausted Brazil and her Argentine and Monte Videan allies, some idea may be formed of the intrinsic wealth and vast resources of the small but prolific province, whose total revenue, under the more ruinous dishonesty of her subsequent government, has dwindled down to an amount often less than £2,500 a month.

Asuncion was so jealously isolated from the outer world that very few travellers have visited it, and there are no hotels. Some palatial residences were built in the town by Lopez for Mrs. Lynch and other members of his family, and one of them now does duty as an hotel, under the title of the "Tramway Restaurant," to avoid the licence that would be exacted from an hotel; it is a spacious, fine building, with a handsome courtyard, but like all the other buildings sadly neglected and dilapidated, and miserably maintained. Our bedroom was a large lofty room, paved with brick, with no furniture except two beds, two chairs, a deal table and a water jug, and hooks in the walls for our hammocks.

There are in the town some very fine specimens of the massive architecture of the early Spaniards and Jesuits. The roads are beds of sand; the only available conveyance was the jolting bullock cart; the men all travel on horseback, and horses are abundant and good, but the native women bring their goods to market on foot, frequently travelling thirty or forty miles barefooted, with heavy loads gracefully borne on their heads, the proportion of women to men being as eight to one. All work is done by women; every morning, and all day, the spacious marketplace was crowded with hundreds of women and girls, each seated on the ground under a full tropical sun, folded in a clean white linen sheet, with her little stock of produce—seldom of the value of a single shilling—spread out on the sand; nothing can be more picturesque than these market scenes; by moonlight they reminded us forcibly of the nun scene in "Roberto." Their produce consists principally of heaps of oranges—not fetching more than sixpence per hundred—piles of mandioca root and dried mandioca, lard, bananas, potatoes, yams, bundles of sugar-cane, batatas, eggs, bamboo canes, tough cheese, tobacco and cigars, maize, bread, gourds, and various other fruits. Every one of them smokes a roughly-made black cigar, and lives on oranges and bananas, mandioca and sugar-cane.

The railway enables them to bring their goods from long distances at very low fares, but as each passenger invariably brings the full weight of luggage authorized, the weighing is carefully attended to by the railway officials. They are crowded together like cattle in the vans, but nothing can exceed their lively cheerfulness and apparent happiness. It appears surprising that there are no general dealers or commission agents to collect these goods and avoid this apparent waste of time, but time is of little value here, and the incessant plunder of which they have been the victims has destroyed all confidence, and the only means by which these poor creatures can profit by their little store, is to bring it on their own heads and watch over it till it is sold. The almost total absence of coin in the country affords another fertile field of extortion; the tramway tokens are largely used as a convenient substitute, but the bulk of their goods is bartered, and constant exchanges are taking place all day long in the market itself. The sandstone hills around the town are covered with a gorgeous subtropical vegetation, interspersed with huts and little gardens, thickly peopled with half-naked, dark-coloured women and children; the latter are absolutely naked.

Our principal object was to visit the interior of the country, and after staying a very few days at the capital, we determined to proceed at once to Paraguari, the terminus of the railway, and forty-six miles from Asuncion. The manager, Mr. Hume, kindly undertook to bespeak the best accommodation he could obtain, and we are deeply indebted to him for the kind reception we met with; but in a country which no ladies had visited, we were naturally not without misgiving as to the hardship we knew they must encounter. The railway rises gradually through a lovely picturesque valley, with seven intermediate stations.

It is flanked on the north by lofty sandstone hills, rising in precipitous slopes to a height of 600 and 1,000 feet, densely wooded up to the summit. The line skirts a large and picturesque lake, upwards of ten miles long, at the base of the hills. It was on the slopes of these hills that Lopez formed his great military camp, consisting of 30,000 men, thoroughly drilled and equipped, and supplied by means of the railway.

The valleys are extremely picturesque and

beautiful. With the exception of groups of palms and orange trees, and other prominent tropical plants, they resemble luxuriant English commons, with numberless copses, and bushes, and pools, and streams of sparkling water meandering among brilliant green meadows through a soil of inexhaustible richness, enamelled with calladiums and other moisture-loving plants and flowers; and the great timber trees, of which the forests behind consist, are themselves, to a great extent, flowering plants of great beauty, covered with brilliant bloom. The hills to the south of the valley are nearer the railway, and from 400 to 600 feet high, wooded at the summit, and at the base forming a continuous series of cultivated gardens, with villages surrounded by groves of oranges, bananas, and palms, interspersed with plantations. The women with their white cloaks, and always bare-footed, crowded round the carriages at every station, offering for sale roast chickens, sausages, cakes, bread, and oranges. Many of the girls are extremely handsome, and move about, all barefooted, with their heavy water pitchers on their heads, with that inimitable ease and grace and erect bearing peculiar to the descendants of the Spanish race, while nothing can exceed their respectful civility and kindness. There were a few cases of goitre and a few disfigured cripples at some of the stations endeavouring to excite attention by their deformity; with this exception we never met with a single beggar anywhere in the interior, nor did we ever come across a drunken man or woman.

The only drink sold at the stations is the canna, or white rum, distilled from the molasses or the cane in the sugar plantations. It is a strong spirit, and though very cheap, and always drunk neat, it is never taken in excess.

This is in strong contrast with the Indians in the Chaco, who sacrifice everything for spirits, and are soon decimated, whenever it is obtainable. A colony of the Indians from the Chaco were allowed to establish themselves on the beach at Asuncion, and held one of their orgies while we were there, beginning with dancing, culminating in a drunken free fight, of the most revolting description.

A very comfortable old first-class carriage had been put into running state and was placed at our disposal for the journey, and although the speed was very slow, we travelled in great comfort. The line, after being totally neglected for many years, had been run until the last engine and carriage would no longer perform the journey in a day. The permanent way, and the massive hard wood sleepers, had lasted fifteen years without renewal, and many years without maintenance. The boiler tubes were nearly all plugged. The boilers were all encrusted, and the steam escaped at every joint, and it was only under excessive firing, and very dangerous pressure, that any progress was made. Stoppages for repairs of both the engine and carriage were of constant occurrence, and as the Government was either unwilling or unable to incur any outlay, even after the loan of £3,000,000, this valuable property, with its very large equipment of old rolling stock machinery and workshops, was transferred for a mere nominal consideration to a few individuals in the town. Under the able direction of their manager, Mr. Hume, two engines and a few carriages were soon put into thorough repair, the stations were rendered inhabitable, and the line and timber bridges put into safe condition, and when we arrived well-filled daily trains had already begun to run again, with punctuality, safety, and profit. The gauge is five feet six inches, labour is cheap and plentiful, balks of hard timber are attainable at 1s. 6d. per cubic foot, hard wood sleepers at 3s. Excellent bricks are equally cheap, and the earth works are finished five miles beyond Paraguari. The completion of such a line, through such a wealthy district, is only one among the numerous inducements that would speedily attract the capitalist under a government with any semblance of honesty or stability.

## CHAPTER XVIII.

### THE COTTAGE AT PARAGUARI.

THE house Mr. Hume had procured for us at Paraguari was kept by an Italian, with a very industrious and intelligent wife, and four or five children. Like all the houses that surround the large Piazza at Paraguari, it was a one-storied cottage, substantially built of mud and brick and hard wood, roofed with Dutch tiles set solidly in mud, on bamboo rafters, with a substantial brick paved balcony in front, and with a shed and a little yard behind, surrounded by a high bamboo fence; we secured a little privacy by tacking up some sheets between the rooms, but as the windows were mere openings with wooden bars,—for there is no such thing as a pane of glass at Paraguari,—and the wind at night is generally very strong, we were often defeated in maintaining this flimsy barricade. Our host had provided a Spanish cook, and a very intelligent Frenchman as guide, interpreter, and general servant. He had seven or eight horses always ready in the stables, and the half-omnibus which plied between the village and the station was taken off from that service whenever we required to use it, and in fact everything that the village furnished was most kindly placed at our disposal; while every alternate day the train brought such luxuries and necessaries as could be obtained from Asuncion.

The population of the district is about two thousand; the better houses, like our own, surround a large piazza. The Spanish villages always consist of a large square surrounded with buildings; in time of trouble the cattle were all drawn into this square, and were easily defended. The mud huts of the Indian natives are dotted round about among thickets of guavas and bromelias. The marvellous contrast between the town and the country, and indeed between Paraguay and other American provinces, was at once strikingly apparent, not only as regards the amiable character of the population, but also in the more material matters of our comforts—cuisine and attendance. Our simple fare was excellent in every respect, with the widest variety; meat, poultry, eggs, butter, cream, game, fruits and vegetables of all kinds, coffee of delicious quality, and good cigars, all produced in the cottages round about us, were supplied in great profusion, and all was well cooked and served.

Although endowed with the highest courage and devotion, as evidenced by the heroic resistance so long maintained against a whole continent in arms, the Paraguayan never carries a weapon of any kind. The swaggering mounted gaucho, with his long knife girdled round his loins, is not tolerated here; animals are kindly and considerately treated, and we felt at once we were living amongst a simple agricultural people of the most gentle and hospitable character. With the most respectful attention, there was no servility; our host considered himself in every respect simply a host, and, with thorough independence, neglected nothing that could add to our comfort. We took part in the evening in the simple harmonies in which his wife and children all joined, regulated by a well-played accordion. We consulted him as to all our movements, and he organized all the little expeditions which we undertook, and occasionally gave us the benefit of his company. He spoke French fluently, and as my knowledge of Spanish was very superficial, and I knew nothing of Guarani, which is the ordinary language of the natives, his assistance was very valuable. Another of my authorities was a very intelligent Frenchman who kept a store in the square, where he was doing a large business and accumulating capital. I frequently called in and had an instructive chat, and a glass of canna and water, and an excellent Paraguarri cigar. The cigars invariably smoked by the women are extremely cheap, but they are badly made, although the tobacco is good; but we found out a Spaniard who had made cigars in Havanna, and thoroughly understood the subject, and from him we obtained cheap cigars of very superior quality, that would command a high price in any European market. But my most important and trustworthy informant was the Gefe Politico, or Commandante, of the district. He lived in a large, massive, and imposing pile of buildings, erected by the Jesuits, adjoining a dilapidated church. It probably formed a portion of one of the great educational establishments with which the early Jesuits have endowed the country. Some of the large rooms were still occupied as schools, for everyone learns to read and write. I was much interested in visiting these schools, and the buzzing sound of a hundred juvenile voices repeating their lessons in union was always welcome music in the square. The education was of course purely secular, for there are no clergy, all Church establishment having been entirely rooted out, and all religious ceremonies, including marriage itself, totally discouraged and almost abandoned. Lopez, for political purposes, did at one time nominate one of his own creatures as bishop, and as he left the Pope no alternative but total repudiation of his authority or his consent to the nomination, it was ultimately sanctioned. He never; however, permitted the exercise of any ecclesiastical jurisdiction. Without any definite creed or any religious convictions, it is still the constant habit in almost every family to assemble in the evening and join in a very short but truly sincere and impressive extempore prayer, bearing a striking resemblance in its practical simplicity to the Lord's Prayer, of which it is probably a simple perpetuation.

My friend the Commandante had charge of the schools,—he was, in fact, supreme in everything. He was magistrate and police, municipal mayor, military governor, jailer, assessor, collector and treasurer, for a population of about two thousand people. The municipal government was charming for its simplicity and efficiency. The taxes were collected monthly, and a monthly account of every item of receipt and expenditure was publicly exhibited on the doors of the Prefecture, and critically discussed. Each tax-payer could thus publicly ascertain whether his contribution had been duly acknowledged and ascertain in great detail its appropriation. The payments included a salary of about £150 a year to the Commandante, watchmen, commissaire, school charges, repairs of roads and public fences, maintenance and cleansing of the markets, public square, and Prefecture, and expenses connected with police and magisterial duties. The total expenditure was about £600 a year. No one could receive any payment without this public announcement of its amount and object. If occasionally a few loads of gravel, or the repair of a bamboo fence, was too dearly paid for, the error could not be repeated without serious remonstrance.

It is difficult to imagine any system more simple or more perfect so long as the Commandante was, like my friend, an honourable man. On the other hand, the Commandante is of course appointed by the Government. There is no independence or stability in his position. He is often a total stranger in the district, and he is selected solely on political grounds, or as a reward for political services. He is constantly liable to dismissal; in times of revolution he is bound to side with one side or the other among the rival adventurers in the capital who seek his support, and it but too often becomes his duty to call out these helpless but obedient peasants, and march them away from their homesteads armed with a musket or a pike, to murder each other in some iniquitous brawl of which the object is as unknown to him as it is to them. The utter helplessness of the population under such Commandantes as Francia and Lopez would employ may easily be imagined.

On the other hand, if we could insure an honest, intelligent, and perfect ruler, despotism is confessedly the most perfect form of government that could be desired, and there is no doubt that a large part of the happiness, contentment, and prosperity of the villagers round about us at Paraguari was due to the genial, honest, and upright character of our friend the actual Commandante. As quarrels, drunkenness, theft, vagabondism, and all the ordinary vices of civilization are almost entirely unknown, his judicial functions were of the lightest character. I only witnessed one instance of their display, which I will relate. The institution of marriage is now coming rapidly into vogue amongst the better people, and a very luxuriant wedding feast, with the usual ball in the evening, was given at the cottage of a well-to-do native near Paraguari. The guests, one hundred and fifty to two hundred in number, rode in from all directions, and refreshments were served on a very liberal scale all day and all night. Among the guests was a native of Corrientes settled in Paraguay,—the Corrientino furnishing, as already alluded to, one of the worst types among the various races that people the South American republics. The Corrientino was a wealthy and important man, and was accompanied by several members of his family. He was remarkable as being the only guest armed with the terrible large knife or dagger invariably carried and used by the Argentine gaucho, and so distasteful to the Paraguayan. In the course of the evening, excited by the dancing and the constant draughts of canna and strong Spanish wines served at the feast, he picked a quarrel with a young peasant in the party, and after a little altercation drew his formidable weapon, which he would doubtless have plunged into his unarmed opponent if he had not been seized by the surrounding guests. He was immediately disarmed, his horses and carriage were brought out, his family fetched from among the guests, and the whole of them were, very resolutely but very quietly, sent home about their business. Our friend the Commandante was an invariable guest at every dance and feast, and directed these proceedings. The dancing was resumed, and the event soon forgotten. On the following day, however, the Corrientino was summoned to the Prefecture by the Commandante, and after a severe reprimand he was fined ten pounds for carrying such a weapon, and warned under penalty of expulsion from the district to avoid such another display. The penalty of ten pounds was paid, and credit given for the amount in the usual balance sheet when exhibited at the Prefecture. This simple and striking display of power and justice told largely in mitigation of such absolute and irresponsible authority, but it equally elucidated the true foundation of the marvellous influence of the kind and beneficent early Jesuit, and of the late merciless and bloodthirsty tyrant of whom even yet the peasants speak in fear and trembling. On settling down in our cottage without doors or windows, and with a tolerable assortment of books, thermometers, barometers, microscopes, and other instruments, in addition to the knick-knacks of the ladies, the first thing that struck us was the absolute security with which we could leave these things about or travel anywhere without attendants in a country entirely without police.

Every morning the market and piazza, or square, was occupied by hundreds of native women in their clean white sheets sitting on the ground chatting and laughing and bartering their simple produce and smoking the roughly-made cigars which everyone indulges in, and which are sold at less than a penny a dozen. We could not but envy their apparent happiness, and realize the simple lesson that its truest basis is contentment. The articles they brought for sale to the house consisted of parrots, shawls, handkerchiefs, and collars of beautiful lace peculiar to the district, elaborately ornamented hammocks, and skins of tigers and other animals. They drove a good bargain, but with singular independence, and without importunity.

Every cottager grows and spins enough cotton for his own consumption, and weaves his sheet and clothing. The hammocks, which are all made by hand, are extremely strong, durable, and handsome. The Paraguayan lace is also very beautiful, the art being another remnant of the lessons of the Jesuits. The Paraguayan puzzle-rings, most ingeniously worked in fine gold, originated from the same source. Amongst other relics of these pious pioneers are the sun-dials found in every village. One of the dials stood in a small enclosure in front of the Prefecture, and is an astronomical curiosity. It consists of a solid column of masonry about eight feet high; the dial consists of a plain block of Lapacho wood four inches thick and fourteen inches square, built into the top of the column, and placed precisely in the plane of the equator, i.e., making an angle with the horizon equal to the co-latitude of the village. The gnomon is a plain, solid copper rod, passing through the centre and at right angles to the block, and therefore parallel with the poles of the earth. The hours are thus equal divisions, as on an ordinary clock, deeply cut both in the upper and lower surface, —the upper surface giving the time in the summer months, and the lower surface in winter.

Such a dial is peculiar on account of the extreme simplicity of its construction and adjustment, which requires nothing but a level. The block is adjusted on the level surface of the top of the column on the day of the equinox, by simply placing it in such a position that both surfaces are in shade during the whole day. I found it extremely accurate, giving the time within a minute, although the block has been exposed for forty years to this tropical sun. The date on the dial was 1838. It was a singular coincidence, that during our short stay in Paraguay we witnessed two eclipses both invisible in Europe. The first was an eclipse of the moon, at Asuncion, on the 23rd August, the second was an eclipse of the sun, seen from Paraguari on the 7th September at 8 a.m., the railing round the dial serving as our temporary observatory. We observed one spot on the sun's disc; the body of the moon was intensely black, and very sharply defined, and the times of contact were observable with great precision. The sun being about half obscured, there was a very sensible diminution of light; and both these eclipses attracted the attention of the natives, who have of course no almanacs. In their use of the dial they take no account of the equation of time, and this was indeed the case throughout the continent.

We made daily excursions, either on foot or on horseback or in our half-omnibus, drawn by three small but excellent horses, and we often took long botanizing rambles quite alone, so that we explored the country for many miles round. Wherever we went we were received with the greatest hospitality, aud presented with fruits and flowers. A favourite excursion was to the establishment of a German settler, about ten miles from the village, who owned a considerable area of land. The wide road, with its deep ruts in the sandstone rock and beds of sand, was scarcely passable for a carriage, but the scenery was magnificent. We crossed a series of commons, densely covered with pine-apples, eryngiums, and bromelias, with thickets of large grasses and flowering shrubs, backed by forests of palms and timber trees, which covered the surrounding sandstone hills. We were there in the spring. The great Lapachas, "Tecoma," with no leaves, but covered with gorgeous flowers, were prominent objects of beauty, towering above the other trees in clouds of flower. There are two species, both equally valuable for their timber,—one with brilliant yellow bloom, the other pink. The large bell flowers as they fell carpeted the ground beneath the trees, and attracted myriads of bees and gigantic butterflies and other insects. The durability of the iron timber of this bignonia appears to be almost endless. Old Jesuit buildings, and even posts sawn off close to the ground, are as sound as when cut a century ago.

We passed on our road a continuous line of cottages, with gardens of the most luxuriant descrip-

on the level surface of the top of the column on the day of the equinox, by simply placing it in such a position that both surfaces are in shade during the whole day. I found it extremely accurate, giving the time within a minute, although the block has been exposed for forty years to this tropical sun. The date on the dial was 1838. It was a singular coincidence, that during our short stay in Paraguay we witnessed two eclipses both invisible in Europe. The first was an eclipse of the moon, at Asuncion, on the 23rd August, the second was an eclipse of the sun, seen from Paraguari on the 7th September at 8 a.m., the railing round the dial serving as our temporary observatory. We observed one spot on the sun's disc; the body of the moon was intensely black, and very sharply defined, and the times of contact were observable with great precision. The sun being about half obscured, there was a very sensible diminution of light; and both these eclipses attracted the attention of the natives, who have of course no almanacs. In their use of the dial they take no account of the equation of time, and this was indeed the case throughout the continent.

We made daily excursions, either on foot or on horseback or in our half-omnibus, drawn by three small but excellent horses, and we often took long botanizing rambles quite alone, so that we explored the country for many miles round. Wherever we went we were received with the greatest hospitality, aud presented with fruits and flowers. A favourite excursion was to the establishment of a German settler, about ten miles from the village, who owned a considerable area of land. The wide road, with its deep ruts in the sandstone rock and beds of sand, was scarcely passable for a carriage, but the scenery was magnificent. We crossed a series of commons, densely covered with pine-apples, eryngiums, and bromelias, with thickets of large grasses and flowering shrubs, backed by forests of palms and timber trees, which covered the surrounding sandstone hills. We were there in the spring. The great Lapachas, "Tecoma," with no leaves, but covered with gorgeous flowers, were prominent objects of beauty, towering above the other trees in clouds of flower. There are two species, both equally valuable for their timber,—one with brilliant yellow bloom, the other pink. The large bell flowers as they fell carpeted the ground beneath the trees, and attracted myriads of bees and gigantic butterflies and other insects. The durability of the iron timber of this bignonia appears to be almost endless. Old Jesuit buildings, and even posts sawn off close to the ground, are as sound as when cut a century ago.

We passed on our road a continuous line of cottages, with gardens of the most luxuriant description. Almost every one is a proprietor. The establishment invariably consists of a large, roomy, substantial cottage, built very solidly, with a wide open verandah, and a large space open on two sides in the centre. The walls are of hard wood, timber and bamboo, solidified with sunburnt sandy clay; the floors are paved with brick; the roof consists of Dutch tiles set in clay on bamboo rafters; the cooking is done outside. The hammocks are slung in the airy central space, or under the verandah. The luxury of swinging to and fro in such a dense shade under a tropical sun must be experienced to be understood. The women are either at work in the gardens, or drying the mandioca in the sun on bamboo frames, or pounding it, and working it into cakes; others were spinning their home-grown cotton, or converting their little crops of tobacco into cigars; some were distilling canna from molasses in the most primitive fashion, or packing oranges in banana leaves. The fetching of water from the numberless brilliant springs that abound everywhere is an important duty: it is carried on the head in large jars of red earth, and a porous water-cooler, made of the same material, unglazed, is found in every dwelling. We brought away some of these coolers, on account of their elegance and efficiency their cost was threepence each. The men are mostly on horseback, well dressed, with clean trousers, vests, and a showy poncho over the shoulders, but always without shoes.

The cottage is surrounded by a cultivated enclosure, invariably containing a few palms, large groves of oranges, a few clumps of bananas, with their polished stems and majestic foliage towering high into the air. Plots of vigorous growing tobacco and cotton, a little field of sugar-canes, vegetables, and fruits of all kinds, including beans, potatoes, peas, batatas, gourds, mammoos, maize, rice, and above all large tracts of mandioca, which is the staple food of the country. In moist spots enormous towers of bamboos add to the luxuriance of the shade. The soil is a rich vegetable mould, blended with coarse sand, of unbounded fertility, and always moist from the numerous land springs originating in the surrounding hills. There is no ploughing or digging, it is simply hoed over with a light tool. The orange groves are indescribably beautiful; the enormous trees were bowed down with their delicious golden produce, and the ground beneath them was covered with fallen fruit. They are used for distillation and for feeding pigs and cattle, and are largely consumed by the natives, but have no money value. The orange is not indigenous, but was introduced by the Jesuits; it is now universally distributed, and enormous forest tracks are covered by the wild bitter orange, produced by the seed of the cultivated plant. In addition to their vegetable wealth, nearly all the cottagers are again rearing cattle: milch cows, draught oxen, horses, and pigs, thrive on the pasture lands. The enclosures are fenced in with bamboo. I measured some bamboos purchased by our landlord. They were sixty-eight feet long, and of proportionate girth; their value was the mere carriage by bullock cart, or about 10*d*. each.

We rode for many miles through homesteads of this character, separated by dense forests of timber trees of the most valuable description, and furnishing the most refreshing shade. The botany was gorgeous and inexhaustible, and not confined to tropical plants; one of the most showy flowers is the Paraguay jasmine, with its party-coloured perfumed blossoms, white, and mauve, and blue. We stung our hands with the tree nettles and cactuses and the sharp spines of the coca palms. The only English plants we met with were some fungi, viz., the universal agaricus-procerus, the common mushroom and morel, and some polypori. The woods are full of vegetable products of great economic value, and are not yet explored. A few home uses attracted my notice. Some of the climbing plants are used as ropes. The caraguata, a species of bromelia, with edible fruit precisely like a pine-apple, if the carpels were separate, furnishes a splendid fibre used for cordage.

The bark of the gigantic carupai is used for tanning. The soft wood of the timboo is used for making large troughs, carved out of the solid trunk in one piece, and resists without cracking the full blaze of this burning sun. The leaves and flowers of the bitter orange are distilled to obtain an essential oil, which commands a high price.

On arriving at the establishment of our German friend, we were welcomed, as usual, with the greatest hospitality. He quickly set before us a sumptuous repast, consisting of excellent soup, sausages, and mandioca, poultry, green peas, and omelettes, oranges, and canna or white rum, and deliciously cool spring water, with excellent coffee. He was a producer on a larger scale than usual; his principal products were sugar, canna, maize, tobacco, and mandioca. He also grew batatas, potatoes, and rice. We visited the sugar plantations, in which every one works barefooted, although a venomous dark viper is but too common; the natives are however extremely sharp in avoiding them, and accidents are not frequent. A boy who had been bitten a few days back had comparatively recovered, but another who had been bitten by a scorpion succumbed; fortunately, the more venomous snakes, such as the rattle-snake, do not frequent the plantations. In wet places boas of enormous size and strength are very common, and the brother of our host had great difficulty in extricating himself from the folds of a large snake. The home-made crushing machinery for the sugar-canes was very primitive, consisting of three well-worked hard vertical cylinders, geared into each other with wooden cogs, and turned by beautiful trained oxen. The distillery was equally primitive, the condenser being a worm in a wooden cask; and to my surprise the water was all brought by hand from a spring half a mile distant. The canna is distilled from the molasses, or simply from the crushed canes, fermented in large solid troughs adzed out from the enormous trunk of the timboo. The canna is a fine and genuine spirit, exclusively consumed all over the continent, and I bought thirty gallons of our friend at a very nominal price to bring home, but the duties have made it a costly spirit here.

The tobacco crop was all gathered; I was informed that the first pluck or the heart is the mildest, the fourth or last crop consists of the outside leaves, which are strongest.

The women were planting mandioca; the loose sandy soil is hoed over, the stems of the old plants are cut up into short lengths and dibbed into the moist and prolific soil. The cultivated spots are all in patches, separated by magnificent forests, with a vegetation luxuriant beyond description

We entered some of these forests where tracks had been cut, but the swarms of mosquitoes in the evening were truly formidable. Our host, with half-a-dozen assistants, owned many square miles of this marvellous land, which was capable of supporting a population of many thousands in luxury and plenty. We were strongly pressed to repeat our visit, and, as a temptation, a boar, deer, and tiger shooting excursion was arranged; but as it involved a bivouac all night in one of these mysterious forests, and as my experience with the gun was confined to gentle pheasants and partridges, this treat was reluctantly declined. Every effort would doubtless have been made to have rendered the expedition agreeable, and everyone volunteered to help in the driving of the wood, in which all the real labour consists. Among the guests at our friend the German's, we met his brother, who had established himself in the very middle of these maiden forests, where he carried on the distillation of the 'essential oils from the wild bitter orange, which covers extensive areas. The scent and beauty of these orange forests is indescribable. He insisted on our paying him a visit, which I did with the more pleasure as I found him a highly intelligent guide, and it afforded us a very agreeable excursion. We rode for miles through these wonderful woods in a shady track, and found his little establishment in an open space in the forest formed by the cutting down of half-a-dozen trees of magnificent dimensions, which lay round about. His little stills were made from ordinary flasks with tin tubes, and his very primitive hut was a mere open shed, covered with reeds; with a couple of berths, very much like those in a steam packet. His companions were a young Frenchman and a fine tame boa in a cask. The essential oil was of two kinds, the more valuable was from the leaves, the other from the flowers of the bitter orange. He knew nothing of the purpose for which it is used; a dealer at Asuncion bought it at about £3 for an ordinary bottle; but he believed it to be of much greater value. It was a very slow process, and produced nothing to compensate him for such singular exile. A stream which ran through the bottom of this forest was a favourite haunt for large boas, and he caught a considerable number. Tame specimens of this snake are very common, and he had sent several to Europe. He related to us a singular adventure with respect to a very large boa, which had just been shipped from Asuncion for the Zoological Gardens. He told us he frequently saw snakes that he dare not attack on account of their formidable size; but on one occasion, his brother and another man being with him, he boldly seized a very formidable snake, as it crossed a marshy road, and called to his companions to assist him in its capture. He had seized it as usual by grasping its neck with both hands, and holding the head in the air; but before assistance could arrive his formidable antagonist had managed to make two coils round his right arm, one round the body, and another round the leg, and so secure was the compression that he was quite powerless. His brother's first instinct was to kill the animal with his axe, but to this he courageously objected, being very anxious to procure such a specimen alive. By dint of sheer force the snake was unwound and dragged home to the hut, where it was kept two months in a large cask, resolutely refusing all food. It became torpid and tame; but on one occasion it inflicted a severe bite on the thumb of a stranger. At Asuncion it became very tame, and ate bread and milk, and arrived safely and in good health in Europe. He has seen larger snakes that he dares not attempt to attack.

Encounters with these monsters are not uncommon. Mr. Horrox, the manager of the tramway in Asuncion, told me that when he was travelling on horseback through some high rushes, his peon, who was in front of him, was attacked in a most determined manner by a very formidable snake. As a rule, they are not at all feared by the natives. Snakes of all kinds are very numerous; we came across two or three varieties of coral snake. Monkeys are plentiful, but we only saw two kinds, one with a swelling like a goitre, and the other a much smaller species.

Another excursion we made was to St. John's Cave, so called by the Jesuits. It is approached by a steep climb of about 500 feet through a dense forest, over rocks with boulders of quartz conglomerate excessively hard. All the pebbles were rounded by water action, and they explained the origin of the fertile sandy valley below. These sands are sometimes of great depths; I saw a well eighty feet deep passing entirely through sand. The hill on which the cave is situated is one of the many isolated mountains that occur between the ranges of hills. It was extremely precipitous, especially at the summit, and apparently all sandstone or conglomerate, covered with forest. The cave was a spacious chamber in the face of a vertical rock excessively difficult of access. It was partly artificial, and is said to have been used as a refuge in troublesome times. The road lay among large lagoons of brilliant water, through a vast marsh filled with water plants, and then through a zone of cocoa palms and dwarf shrubs, until we entered the forest that covered the hill. The jaguar, puma, and wild boar are but too plentiful in these forests, but they seldom approach the cultivated districts.

Our village life partly rivalled in interest our wandering expeditions. Dancing appears to be the dominant passion with everyone, and village balls are of weekly occurrence. We had not been many days resident, when we were invited to one of these simple private entertainments; and our friend the Commandante,—who, by-the-by, seldom lost the chance himself,—having advised us that the ladies ought not to refuse, we all agreed to accept the invitation. Poverty is of course totally unknown, but its twin sister, Wealth, is equally a perfect stranger in these happy groves. The ball was therefore of the most simple character, and among about 100 swarthy-looking girls we recognised many whom we had seen sitting all day in the markets, or fetching water from the springs. There was a very disproportionate number of fine lads and men. The room was a very large one, but very sparingly lit up with two or three paraffine lamps. The music consisted of four wind instruments, with an enormous proportion of very monotonous bass. Canna was handed round in tumblers, with tumblers of cold water *ad libitum*. The lasses were all barefooted; their dark shoulders were mantled with brilliantly white and highly-decorated dresses. The men wore high boots, and ponchos on their shoulders. The dances were quadrilles and polkas. The dancing was excellent, and was extremely quiet and graceful; and though a great crowd surrounded the cottage outside, and canna was freely distributed amongst all, no one ever dreams of taking it to excess, and the whole proceedings were conducted with singular propriety, order, and intense enjoyment. We remained a couple of hours, but the dancing was kept up all night. When we returned home about 1 a.m., the night was bitterly cold and bright; the thermometer was 38° under the verandah, and 32° on the grass. This of course is singularly abnormal in the tropics, and we had to abandon our hammocks, and pile all the clothes we could muster on our bed. The date was August 30th, corresponding with March 2nd in England. Only ten days before the night temperature was 90°.

We attended other balls, and especially one given in our honour by the Commandante himself, in his spacious rooms; but the company on this occasion was more select, and the ladies wore shoes. It was at last intimated to us that we were bound to give a ball ourselves. As we had no room large enough in our house a neighbour at once volunteered the use of a spacious apartment, and I instructed my assistant to attend to the arrangements, and omit nothing that the neighbourhood could supply. We were, moreover, honoured by the presence of the manager of the railway and of the tramways, who came from Asuncion. The guests, to the number of fifty, were specially selected, mostly proprietors and tradesmen from many miles round. The music was good and very lively, and consisted of a rude harp and accordion, and a real nigger with castanets and a triangle. The floor on this occasion was carpeted with a Brussels carpet, and the room decorated with two looking-glasses, and some laurels, and lit with lamps and candles. The refreshments consisted of canna and strong Carlone, and a dry Aosta wine, with cakes and oranges. Some of the girls and young men are very handsome. The girls, with their excellent figures and no tight waists, wore loose white dresses, with a little display of lace and colour. The men wore their picturesquely coloured ponchos, with white vests and high boots. The dancing was extremely quiet, with all that natural grace which is so peculiar to the Spanish race. The dances were quadrilles, the lancers, polkas, mazurka and gentle waltzes, with a native measure called the Dove, which was very pretty. The house was as usual surrounded all night by a crowd of half-naked spectators, with dealers in drinks and fruits, and children perfectly naked. As there were no doors or windows, it was not easy to ascertain the limits of our party, more especially as the refreshments were handed as freely to those outside as to those inside; but as usual there was not the slightest disorder or confusion, nor any case of drunkenness or incivility. Some curiosity was naturally excited by my wife and her friend taking part in the first quadrille. Our ignorance of Guarani and Spanish was of course a drawback, but several guests spoke French. Our leader of the ceremonies was a very handsome girl about twenty, the daughter of the landlady of the house. I never witnessed more happiness and real enjoyment, nor more apparent innocence, in any London ball-room, than at this simple entertainment, the whole cost of which, including music, was under thirty shillings. Happy people! Are they to be envied or to be pitied? Most of the men maimed in the late wars, five-sixths of their number buried in the neighbouring woods and marshes, or on the sunburnt plains of Brazil or Corrientes, yet still contented, happy, and peaceful, and totally ignorant of any single grievance. Although unchecked by police, or law, or education, or even religion, thefts, drunkenness, quarrels, and crimes of violence, are almost unknown. Such are the marvellous remnants of Jesuit teaching among this docile race, so specially adapted for its reception. In no part of the world can the traveller wander about with more security and hospitality; nowhere can he find more magnificent scenery nor a finer climate, nor a more prolific unexplored and unknown field for investigation.

It was a natural enquiry of great interest to ascertain the physical characteristics of this little province, which separate it so prominently, like a garden of Eden, from all the sunburnt deserts and uninhabited wastes that surround it.

Paraguay is essentially a region of mountains and springs, definitely limited, like an island, by the great rivers and marshes which enclose it, and into which its numberless little rivers and rivulets drain in every direction. Ranges of mountains and isolated hills rise on all sides out of a series of sandy plains, forming innumerable valleys. The mountains and hills are composed of decomposing sandstones, with pebbles and quartz conglomerates of very ancient geological date, densely covered with forests of marvellous luxuriance, and deep beds of vegetable mould. Its temperate climate is due to this superabundance of vegetation. The temperature is lowered by evaporation, and the mountains maintain in a tropical atmosphere a cooling mantle of cloud. The condensation of this moisture supplies the endless springs and streams that meander gently in all directions through a porous sandy soil towards the surrounding rivers.

The tropical storms that occasionally devastate these mountain forests bring down these fertile decomposing sands, mixed with solid streams of the richest vegetable mould, and distribute them in deep beds far and wide over the valleys, but principally along the singularly fertile zones that are found at the base of every hill. These zones are always distinctly marked out by the stately cocoa palms that flourish in such a soil, and form broad and continuous belts surrounding every valley. These are the spots selected for the numerous homesteads and gardens, of which a description has been given. The clearing consists in removing the brush and underwood, and so many of the palms as come directly within the limits cultivated. The remainder are preserved on account of their valuable produce.

In the wars of Lopez, thousands of men and women were driven into the woods, living almost wholly on the great bunches of fruit so abundantly produced by those lofty and graceful plants.

The inexhaustible fertility of such a soil under a tropical sun, permeated everywhere by springs, and refreshed by copious dews and rains, may easily be imagined. This fertility, moreover, is not limited to the hillsides; the whole of the valleys are carpeted with pasture and luxuriant herbage, watered by streams and marshes and lagoons, sometimes assuming the proportions of vast lakes. That such a country should abound in animal as well as vegetable life in all its forms, is only natural. The woods and waters teem with wild animals, and birds, and fish, and reptiles; and the sportsman may reap as rich a harvest as the naturalist.

The climate of Paraguay would well reward careful investigation. It is, as a rule, temperate and fine. The northerly or equatorial winds are not prevalent; but when they do occur in the summer months, these heated waves, though slightly tempered on the hills, produce at Asuncion a temperature as completely tropical as in the centre of Brazil, and the heat is unsupportable; even in September the temperature at Asuncion was 99° for three successive days, falling at night to 85°. In the summer the hot sands become unbearable for the naked feet of the natives, and they are prevented from travelling during the day; but on the hills the nights are tempered by a constant breeze. As a rule, the winds sweep up or down the valley at Paraguari, either from due north or due south; the constancy and strength of a cold current that invariably sets in every evening, was a phenomenon I could never satisfactorily account for. It began as a gentle, pleasant breeze at sunset, preceded by a little haze, and increasing sometimes to the force of a hurricane, shaking the strongest buildings, and fully accounting for the massive strength of the low, one-storied cottages of the natives. Its most striking feature during our stay was its low temperature, rendering the nights so bitterly cold. The thermometer on the night of the ball at 2 a.m. registered 38°, at 2 p.m. it was 63°.

Although coming from the south, it was impossible to conceive that this cold river of air could have passed over the heated rocks of Uruguay and Brazil. It was more probably a current from the higher regions of atmosphere deflected down upon us from local causes, due to the configuration of the great ranges of hills around us. Its force was especially concentrated at the railway station, where a massive building of masonry is with difficulty maintained.

This current is more or less constant throughout the whole year, and materially tempers the climate of Paraguari. We had, however, some experience of the disagreeable character of an equatorial current, which lasted six days, sometimes with the force of a gale, and which strongly reminded us how close we were to the tropics. The temperature, which on the first was 38°, now reached 93°; but whether the winds were north or south, the striking difference of temperature between night and day was equally manifest.

Taking the mean of the maxima and minima for seven days, the night temperature with the north wind was 53°, the day temperature 76°; with the south wind, the night temperature was 46°, the day temperature 70°.

Thus the difference of temperature due to wind alone was only 6°, while the difference of temperature between night and day was no less than 24°, whether the wind was north or south. The south wind was dispersed and driven back by a southern current, with lightning and a deluge of rain as at Buenos Ayres.

This was followed by a delicious change of temperature; but for some hours low layers of cloud hid the summit of the surrounding hills in all directions for many miles, the base still dissolving into very fine rain, while small detached masses of vapour at a lower level drifted by with apparently great velocity. The height of this uniform layer of cloud was 200 feet. The terrestrial layer beneath it retained its transparency, its temperature being maintained above the Dew Point by the warmth of the earth.

During these hot sirocco winds it is impossible in the day time to make any exertion; we could only lie still in our luxurious hammock. A siesta in the middle of the day is the universal habit. Our landlord and all his household disappeared from twelve to four, and I always devoted this interval to a dreamy swing in the hammock, meditating on the beauty and novelty of the surrounding scenes, and on the happiness and contentment of the population.

"What is happiness?" I said, as I kicked off my slippers. The learned discourse of Cicero bears no more relation to the happiness of these simple Indians, than a discussion on transubstantiation does to the devotions of a fire-worshipper. It is far less difficult to theorize as to what happiness ought to consist in, than to determine practically what it actually does consist in, and this element must vary in every individual according to his education, position, temperament, and age. It has often occurred to me, apart from all higher consideration, that the only practical means of definition will be to accept each individual's own standard, which will certainly be the only standard he will acknowledge. For this purpose let us suppose that every human being always carried with him two bags, containing an indefinite number of black and white balls, and let us farther assume that at every emotion of his active life, according to his own appreciation of his material happiness, he throws these balls into a third bag,—a white ball to express happiness and contentment, a black ball to represent unhappiness and discontent; so that if these balls were ultimately counted, they would furnish a correct account of his own estimate of the happiness or misery of his life.

We need not, however, trouble ourselves to count the balls at all, or even to collect them, if, as I believe to be the case, we can, à *priori*, predict the exact result, which will simply be, first, that the number of black balls will be precisely equal to the number of white balls, whatever may be the apparent happiness of the individual to others, and whether he be an Indian or a Newton; secondly, that the total number of balls collected by different individuals will be proportionate to their respective activity and vital energy, whether for good or evil; that is to say, taking an extreme case, a being who, like a vegetable, merely lived without emotions of any kind, would collect no balls at all.

That the number of black and white balls must of necessity be equal, is evident from the following consideration. Enjoyment or happiness in this sense, like length or height, is purely relative, and its actual amount can only be estimated by comparison with some standard, which must in this case be the mean state of happiness or enjoyment of each separate individual. The Indian, or any one of us, would only take the trouble of registering one or more white balls when happier than *usual*, or black balls when less happy than *usual;* but the " usual " must of necessity be the mean state of existence in which alone no balls will be thought of. No other standard can be possible. This mean or usual state of existence may and does vary as regards positive means of enjoyment to any extent with different individuals ; but it will itself be always mathematically defined for any given period by the actual equality of the number of black and white balls collected during that period.

If this evident postulate be admitted, it leads to the highly probable conclusion that all men enjoy a precisely equal amount of happiness even in their own estimation, and totally independent of their wealth or health or ignorance, or even their failings and their vices. The mean state alone varies in different individuals; but the mean state is absolutely special. It is the standard each must use, and in no way affects any other individual.

The black and white ball theory is equally applicable to every single action of our lives ; there is no pleasure without its concomitant drawback, no waves of happiness without corresponding depressions in every phase of our existence. Leverrier half filled his sack with white balls when it was announced to him that his new planet was discovered ; but the other sack was already nearly full of black balls, dropped in during the long vigils and labours that led to the discovery. The school-boy's holiday balls precisely balance all his quarter's woes and cares. The gambler fills his bag at every lucky turn of fortune. Even the suffering patient, on a bed of sickness, adapts his mean to his position, and throws in white balls when pain ceases, or when again allowed to enjoy the cheering sunshine in his garden chair. The mean is sometimes so humble that a mutton chop bestowed on a starving pauper fills his happy store. With another the mean is so inaccessible, and his sources of happiness so limited, that nothing but a Batoum or a Kars will excite a single dip into the bag. Among the affluent and powerful, the actual sources of happiness are dwarfed by the magnitude of their standard in precisely the same ratio as they are enlarged by the extent of their resources. And thus every one will be found to be precisely equally weighted, right and left, whenever he sits down to count up, even according to his own estimate, the happiness or misery that has fallen to his lot. The high or low mean has given no advantage or disadvantage. The Guarani Indian, the invalid, the pauper, and the emperor, may rest assured that neither has been favoured nor neglected, but that as regards the abstract enjoyment of life, or material happiness, it has been bestowed upon all with absolute impartiality.

But what inducement is there for good or evil in such a philosophy? There is certainly no inducement to do wrong, for if we are fully persuaded that the two or three white balls that any indulgence may afford will be accurately balanced by a corresponding number of black ones, the temptation to evil will be completely removed.

On the other hand, we are endowed with absolute freedom to do right, and with every encouragement to climb higher and higher, rejoicing in the certainty that the black balls of labour and self-denial are but stores of happiness which "no rust nor moth can corrupt, and no thief break through and steal."

As sleep crept over me in these hammock meditations, and fancy took some wilder flights, I found I trenched on failing ground. Do angels have a mean? or is its presence not the rigid test that separates the finite from the infinite, the cloud that shrouds infinity? Time without beginning or end can have no mean, no earthly standard for its measure,—space eternity, etc.

I was here fortunately awoke. I had clenched my argument by thrusting my leg through the mosquito net, and a savage gnat had instantly plunged his poncho dagger into the flesh, and properly recalled me to my editorial duties.

The traveller in the tropics must be provided with some black balls even at this time of the year. The enjoyment of the siesta is materially balanced by the determined warfare of the mosquitoes, although they are far less numerous in the open village than in the valleys and woods. They can't stand the wind. We managed to guard against them with our nets even when sleeping in the hammocks. Oil kept away the harvest bugs, and insect powder bothered the fleas and other domestic pests; but our worst enemies were the jiggers in our feet, which no precaution could keep away. The jigger is a species of flea that probably breeds in the sandy floors; it is excessively minute, and buries itself beneath the skin so cautiously, and with such perfect absence of irritation or itching, that it is impossible to become aware of its presence except by seeing it, and this unfortunately is seldom done before the bag of eggs it has deposited has attained considerable dimensions, and created a corresponding cavity in the flesh. If neglected, the larvæ rapidly bury themselves in the flesh, and make a dangerous sore. The feet of many of the natives are sadly crippled by these pests. Our only protection was to have our feet examined every day by a negro girl in the house, who had a marvellous knack of seeing them and extracting the little bag of eggs with a penknife; but they occasionally escaped her vigilant search.

After our daily rambles the delicious evenings were always occupied either with the microscope, in examining the natural treasures we had brought home, in rambling round the square among the natives, in listening to the singing of our host and his family, or in watching some motley throng that had collected round the village ball; and it was with sincere regret that at the call of sterner duties we at length bade adieu to these simple scenes and left the lovely gardens of Paraguari.

On returning to Asuncion a hot sirocco wind was blowing over the burning sands. The temperature of the densest shade was 99°, of the river itself 79°, and the heat was accumulated and concentrated in the terrestrial layer by a haze, due to the saturated layers above us. Our steamer, the *Taraqui*, left the quay September 20th, at 6 p.m. The heat and mosquitoes of that terrible night will not easily be forgotten. At 10 p.m., with a temperature of 90°, the steamer stopped opposite the little village of Ipani, to take on board no less in number than 250,000, or, in weight, three tons of oranges. These orange groves, which cover the hills two miles from the river, had been planted by the Jesuits on account of the facilities offered for exportation. The fruit had been gathered and brought down to the shore, and now covered some acres of the sandy river beach in golden heaps on the bare ground. A temporary stage of bamboos and timber had been erected, projecting about forty yards into the deeper water, but in coming alongside our clumsy navigators managed to ground the steamer on some rocks. The vessel had been prepared for receiving the cargo by the construction of large pens on the deck, fenced round with bamboos; about five feet high, and occupying nearly the whole available space, leaving only a narrow passage round the pens for access to the cabins. The night was spent in the noisy operation of hauling off the vessel by means of anchors and capstans, and in taking this fragrant cargo on board. No sleep was possible, and we lay panting on deck watching the process. When at last the vessel was brought alongside, a continuous column, of fifty or sixty women and girls, with the most perfect discipline, each bearing a large pannier of oranges on her head, walked along one plank of the jetty, tipping her cargo of fruit into the pens on board, and returning by another plank.

This process occupied the whole night, and a singular scene it furnished, but not without a few black balls as far as we were concerned. These oranges were destined for Buenos Ayres and the surrounding camp, and it is considered a successful venture if 60 per cent. of them arrive at their destination in good condition. A very little care in packing would avoid this terrible waste, not as regards the oranges, for they have no value, but as regards the tonnage of the cargo. We left at 6 a.m.

the following morning, with a marvellous change of temperature. A polar current suddenly swept away the heated atmosphere, and in a few minutes the temperature fell from 98° to 77°, and in a few hours to 60°. And this with only a little haze, but no cloud or rain.

At Pillar, on the following day, though so heavily laden, our captain's cupidity unfortunately induced him to take in tow a large bark containing no less than a million oranges, for we were in the height of the harvest. This heavily-laden vessel was lashed alongside us, and a strong head wind having now increased greatly in force, our progress was sadly impeded, and the spray blowing up between the two vessels deluged our decks and oranges, and confined us to the cabins, which were also closed. At Pillar the beach was also covered with a golden harvest of oranges, and many vessels were taking in their cargo. At Corrientes the whole night was again spent in uproar, and in still further encumbering the decks with cargo, principally skins and several hundred bags of matté sewed up in skins, which being green when used, contract in drying, and make solid packages for exportation. Many third-class passengers came on board, and we were now truly miserable and overloaded, although we had left behind us cargo sufficient for half-a-dozen more steamers. The river was low and the navigation difficult, and although we discharged a few matté bags at Goyaz, we shortly after drifted helplessly on to a bank, from which no efforts could release us. We appeared to be in the middle of a lake, in 8 feet of water, which was the draught of our steamer, and in a rapid current. Our companion drifted away and carried away our bowsprit. The crew worked all night laying down chain cables and anchors, which were all broken by our steam capstan, thus totally preventing all sleep. These efforts occupied all the next day, and it was not till 7 p.m. on the third day that we were again afloat. The weather was now cold and cheerless, and after other mishaps we were only too glad to get on shore at last at Rosario, after a very disagreeable journey that had lasted no less than eight days.

## CHAPTER XIX.

### CORDOVA AND THE SIERRAS.

THE temperature at Rosario with a gale of wind was 45°, and was very trying. We stopped here for the purpose of visiting Cordova and the Sierras. We are under deep obligations to the manager of the Central Argentine Railway for the very luxurious manner in which we were enabled to make the journey from Rosario to Cordova, a special carriage being placed at our disposal for the purpose. The journey occupies a day, the distance being 246 miles. The railway traverses an ocean of pampas in an absolutely straight line, and apparently so level as to furnish a fair horizon all round, as though at sea; but the pampas in reality rise as we approach the Sierras, so that Cordova is no less than 1,250 feet above the river. At Rosario, the change of soil and climate as we rise is very marked; near Rosario the normal alluvial soil is fertile, and large areas were under cultivation, producing wheat and maize; the crops, however, are subject to destruction by droughts and locusts and storms. The alluvium gradually disappears, and we traversed enormous areas of dry white clay, which was raised in clouds of dust by the strong winds that always prevail. The ranchos are poor and miserable, and the country is covered with scrubby woods and low trees, with a total absence of rivers or water.

It assumes still more the character of a wilderness as we proceed, and we next pass over miles of barren sands and pebbles with scrubby underwood, till we merge at length into the white and dry and dusty rocky district that surrounds Cordova. The climate here is dry, fine, and sub-tropical, and the country is covered with underwood and coarse grasses; the handsome leafless yellow retama flourishes everywhere, and tillandsias and aloes and cactuses cover the rocks.

Cordova itself lies in a valley, and the view from the heights, on approaching it, with the Sierras in the background, is very fine. The "Primero," now nearly dry, runs through the valley, and supplies the town with water.

Cordova owes all its importance to the Jesuits. They founded and erected the town, with its numerous fine churches, convents, and other ecclesiastical buildings, and its massive Moorish cathedral. They also founded the university in 1613. They are still numerous and influential; but the university having now become a purely secular institution supported by the national government, with German professors and teachers, their power is seriously shattered. The fruits of their withering influence, which, unlike Paraguay, has here been unchecked, are but too evident in the listless, bigoted, and superstitious character of the population.

We witnessed a religious procession in honour of a Saint Francisco, the patron saint of the city. The saint and some female figure were paraded round the town in the form of hideous colossal statues, followed by a host of clergy, dressed in gorgeous colours and gold, with military bands of music. Temporary altars were erected in the streets, and it was singular to witness the apparently sincere but abject devotion with which the crowd, consisting principally of women and children, knelt down and crossed themselves as this tawdry procession passed by; and, indeed, no spectator could safely remain uncovered. The maintenance of such an army of priests must prove a terrible tax on the earnings of these poor creatures, by whom they are almost wholly supported. The altars were covered with wax candles, some of enormous size, and many were carried in the procession. The revenue derived from wax alone must be very considerable; but no other material could have been selected so especially adapted for religious offerings. It is remelted without injury, and the same wax may profitably do duty over and over again in a hundred atonements.

A peculiar feature at Cordova is a large artificial lake or reservoir, surrounded by a promenade, well shaded with trees. The heat in summer is intense, and all who can do so retire to the Sierras, but this promenade forms an agreeable evening retreat for the townspeople. A clear stream of water runs daily for several hours down the principal street, from which the roads are watered and many of the houses supplied. The water supply is nevertheless precarious; the river was reduced to a small bright stream, running among the granite boulders and pebbles, which gave evidence of its occasional volume and velocity. It comes down from the Sierras as a mountain torrent in wet weather, and is difficult to control.

Cultivation in these arid valleys, where rain is often totally unknown for many months together, can only be carried on by means of irrigation, and is therefore confined to the neighbourhood of streams. The Primero is in this way robbed of its waters, which are diverted by numberless channels, and absorbed by the fertile belt that lines its course. The produce and trade of Cordova is steady but small; but it will doubtless be improved by the prolongation of the railway from Cordova to Tucuman, a distance of 337 miles. This line, though normally completed, is in so sad a state, and so insufficiently provided with stock, that it is not as yet available for extensive traffic; but the cultivation of the rich and extensive districts around Tucuman, which are adapted for the growth of every kind of tropical produce, and the completion of this railway, will not fail to create a movement that will soon be felt from Tucuman to the Plate. Bonded custom houses are already authorized, both at Rosario and at Campana, and the traffic on the Central Argentine, even under present difficulties, is rapidly increasing.

A large proportion of our enjoyment at Cordova was due to the kindness and hospitality of our friends,—Mr. Gould, the astronomer at the observatory at Cordova, and the professors at the university, who were our constant instructors and companions. The library and fine collection of plants at the university, the museum, the laboratories, and more especially the valuable local knowledge and personal experience of the professors, were always at our disposal. The university, by furnishing a centre and home for science, has been of the highest value in these new countries, where it finds no other support or encouragement. The buildings and cloisters are handsome and commodious, and are still being enlarged. The staff is of the highest order, but its advantages are not yet utilized to the extent that might have been expected. The museum contains an admirably arranged collection of the rocks and botany of the neighbourhood, and the nucleus of a fine natural history collection. A very valuable and extensive collection of dried specimens of plants was in course of preparation by Mr. Hieronymus, for the exhibition in Paris. The eminent services rendered by that indefatigable botanist are well known, and during my whole journey I enjoyed nothing so much as our botanizing expedition with that gentleman into the Sierras. He possessed some magnificent mules, and the equipment of his intelligent attendant, with his large leather botanizing bags slung on the saddle, was complete. We made it partly a picnic, as we were accompanied by Mrs. Gould and other ladies; and we roasted a large portion of an ox that we had brought in the bags, in true native fashion, on a stick over a bonfire. The ride over the sunburnt plains, yellow with the luxurious retama and various dwarf acacias, and the ascent among the hills, which were covered with luxuriant vegetation, were truly enjoyable. We bivouacked in a shaded glen on the banks of a brilliant stream of cold water, sometimes collected into large clear basins scooped out of the granite rocks, at other times tumbling over little falls or gurgling among the pebbles and great boulders brought down in its wilder moods; but the feature which it is impossible to describe was the luxuriance and novelty of the botany that covered every rock, and often the stream itself. The enjoyment was tenfold increased by the pleasure of such a botanist as our friend, who thoroughly knew the locality of every plant for miles around. The tillandsias are especially numerous and beautiful. There are twenty or thirty species, and I was struck with the agility and energy with which our friend would suddenly disappear high among rocks and woods totally inaccessible to me, returning in a few minutes with any particular species to which he wished to draw my attention.

Another excursion we made was to a picturesque mill in the Sierras, on the river Tercera. The route lay for twelve miles over a dusty, sunburnt, hard, white clay, with boulders and beds of water-worn pebbles, and blocks of primary rock with micaceous sands. The heat in the sun was very trying. The irrigated enclosures afforded some shade, as they are surrounded by tall poplars and willows. The grass is almost confined to one single species, the very long fine paspaeum notatum, which grows in patches and feeds numerous goats, and a few herds of cattle and horses. The river runs under a high baranca of sands and pebbly clays; the water was low, there had been no rain for seven months. We met here, as everywhere else, with the most unbounded liberality, and enjoyed a truly sumptuous repast, prepared for us by Mr. Taylor, the proprietor. We also witnessed here the lassoing of some hundred cattle in an enclosure, for the purpose of marking them, and the extraordinary skill of the poncho was very remarkable, although we could not but feel pity for the animals, which, when lassoed at a full gallop, are thrown so violently down on the hard ground as occasionally even to break their backs.

The prominent plants found in the Sierras were the curious azola, growing in the streams, the tillandsias, the zanthoxyllon, the beautiful cestrums, the retama, cassia aphylla, which covers the arid clays with yellow flowers, several clematis and other climbing plants, the budleias, the various species of baccharis, and some gigantic tree-like cactuses, "cereus," which were very striking objects. Among species totally new, Mr. Hieronymus showed me a very curious species of gnetum, found on the Algarobo, and the singular water plant lillæa subulata, —a new genus, necessitating a new national order.

We passed also some extremely pleasant hours at the observatory with Mr. Gould. The building is on a dry and very dusty eminence 100 feet above the town; all around it the white clay is ploughed out into deep ravines by the rains in the most singular manner. The ground is so cut up as to be impassable, and it looks as though it had just emerged from a stormy sea. The observatory is a spacious and commodious establishment, though entirely free from ostentation. The main object with this indefatigable astronomer has been the supply of a want long experienced, viz., the publication of a perfect catalogue of the stars in the southern hemisphere. The principal instruments were specially designed for that purpose. They consist of a large meridian circle and a fine equatorial by Alvan Clark, with the usual accessories, and a collection of standard meteorological instruments, as this department also falls within the province of the astronomer. With the assistance of an efficient staff of assistants, the requisite routine observations have been carried on for some years with a zeal and continuity and completeness probably unrivalled, and the important result of their labour is in course of reduction and publication. The care and undisturbed application involved in the astronomical work is more than enough, without the wearying detail incident to meteorology, and it would surely be a welcome relief and an advantage to both departments if they were kept separate.

Two extensive and instructive volumes of meteorological observations, compiled at the observatory, have just been issued. They possess great interest, as being the first publication of the kind in this climate. Among the graphic illustrations there is a plate which demonstrates in the most striking manner the connection between sun-spots and temperature.

The site for the observatory was selected by Mr. Gould himself. The dry climate and clear sky were important considerations, but the complete seclusion afforded by a position so remote and inaccessible from every busy centre of civilization had, as he informed me himself, no small share in the selection.

In other respects, it would have been more convenient if a public establishment of this kind had been erected nearer the capital. The climate is all that could be desired at Buenos Ayres itself, where the absence of an observatory is a want really felt, although it is connected by telegraph with Cordova. We regretted very much that we could not make a longer stay at Cordova to join in wider excursions among the magnificent ranges of hills, or sierras, which almost come down to the town, and which rise in higher and higher ranges of mountains until lost in the great chain of the Andes, of which they are the outlying sentinels. It is usual to extend these excursions over two or three weeks, travelling on horseback, and carrying tents and provisions for the journey, sleeping at night in the tents. The unwillingness with which I was obliged to refuse the invitation of Mr. Hieronymus to accompany him on such an interesting expedition may well be imagined. The heat is a drawback in the valleys, but this is soon left behind, and among the hills it is only from the cold at night that any real inconvenience is experienced.

We bade adieu with regret to our many excellent friends at Cordova on the 10th of October, and returned to our old quarters at Buenos Ayres, bringing with us several hundred dried specimens of plants, a puncheon of canna, water coolers, hammocks, lace, bottled snakes, geological specimens, and a couple of parrots, with all of which we returned to England in December 1877.

# APPENDIX.

## ESTIMATION OF HEIGHTS BY THE BAROMETER.

HIS universal and invaluable instrument is seldom thoroughly understood, and its corrections are often totally neglected or misapplied. The aneroid records, without correction, the actual weight of the superincumbent atmosphere precisely, as though it were weighed in a spring balance. The pressure of the atmosphere causes the elastic sides of a metallic box, from which the air has been exhausted, to collapse, for it is strong enough not to be entirely crushed; and the extent of this collapse is a measure of the relative pressure of the atmosphere, and is recorded on the dial. The equivalent in inches of mercury is ascertained by experiment under the receiver of an air pump. It is generally adjusted to agree with the mercurial barometer when the mercury is at temperature 32°. It is not therefore an independent instrument, but when properly adjusted and carefully used, the error, with a good aneroid, will not exceed $\frac{3}{100}$ of an inch, and it will indicate a difference of level between a table and the floor. I have two aneroids made

by Elliott, that have not differed $\frac{2}{100}$ of an inch during two years, when compared at uniform temperature.

The height of the column of mercury in the barometer is directly proportional to the pressure or weight of the atmosphere above it.

The absolute height of the atmosphere is unknown. It does not probably exceed 50 miles.

As we ascend the barometer falls, but not simply in proportion to our ascent. And the question arises, What is the corresponding difference of level due to any given fall in the barometer?

The column of air above us, whose weight is given by the barometer, is not like a column of water, or mercury, or bricks, in which the weight at any height is simply proportional to the height above, but on account of the compressibility of the air a cubic inch weighs more near the earth than at a higher level, *in proportion to its compression*. The column of air thus resembles a column of bricks, in which every brick is heavier than the one above in a definite proportion. Therefore, as we ascend, the pressure diminishes in a twofold ratio; not only are there less bricks as before, but the bricks are individually of less and less weight.

If, therefore, the heights are represented by an arithmetical series, the corresponding pressures will form a decreasing geometrical series; *i.e.*, if the heights are logarithms, the pressures will be corresponding natural numbers.

Since the barometer readings will be proportional to the pressures, the actual difference of level between any two stations will be given by the formula

$$\text{Differ} := log. \frac{H}{h} \times c.$$

# Barometric Heights and Corrections. 351

When $H$ is the barometric reading at the lower station, $h$ is the barometric reading at the upper station, and $c$ is a factor, depending on the density or temperature of the air and some other minor influences which are too small to require consideration. In latitude 52°, with mercury at 32°, and when the temperature of the air is 32°, $c = 60158\cdot6$.*

For the difference of level due to a barometric rise of one inch, we have therefore

$H$ at sea level, say 30 inches.
$h$ at altitude required, 29 inch.
$$\therefore a = log. \frac{30}{29} \; 60158\cdot6 \text{ feet.}$$
$$= \cdot01472 \times 60158\cdot6 = 886 \text{ feet.}$$

at which height the barometer will have fallen 1 inch. 886 feet is the value of one inch at the mean barometric height—29·5 inches.

It is of course essential that the temperature of the water or mercury should be precisely the same at both levels; for a barometer with hot mercury or hot water is simply a different instrument from a barometer with cold mercury or cold water, and their readings are not comparable.

All observations are usually reduced to temperature 32°. In order to reduce barometric observations made with mercury at any temperature to what they would have been if the mercury had been at 32°. Since

---

\* The factor $c$ is the height of the homogeneous atmosphere multiplied by 2·3026 to convert Napierian into common logarithms. The height of the column of homogeneous atmosphere that balances 30 inches of Mercury, both at temperature 32, is 26126 feet.

Density of Mercury is 10500 times that of air.

mercury expands in volume $\frac{1}{10000}$ of its bulk for every increase of temperature of one degree, and when confined in a tube this expansion is wholly in length, the correction will be simply the difference of temperature multiplied by $\frac{1}{10000}$ part of the total height of the given column.

If the barometer reads 30 inches with mercury at temperature 60°, then for the reading with the mercury at 32° we have difference of temperature 18°,

and $18 \times \frac{30}{10000}$ or $18 \times ·003 = ·084$,
and $30 - ·084 = 29·916 =$ the corresponding reading when the mercury is at 32°.

Since the barometer at ordinary level is usually near 30 inches, the correction is always approximately obtained by simply multiplying the difference between 32 and the actual temperature of the mercury by ·003.

Secondly, it must be borne in mind that the constant c is based on the assumption that the temperature of the air is 32°. For any other temperature we have the following correction. Air expands very nearly $\frac{2}{1000}$ of its volume for each degree of increase of temperature. By an increase of temperature of one degree, therefore, the total atmosphere is increased in height at the rate of 2' feet per 1000 feet of elevation, or very approximately at ordinary levels 2 feet for a rise in feet due to 1 inch of the barometer.

Hence this very simple rule:—For any excess of temperature in degrees above 32 add double the excess to the value in feet for 1 inch found as above.

That is for temperature $32° + x°$ add $2x$, and vice versa for temperature below 32°.

Thus for temperature 60° add double 28, and for the

value in feet of 1 inch at temperature 60° with the barometer at 29·5 we get 886 + 56, or 942 feet.*

In this manner the following short and useful table, abbreviated from Col. James, has been calculated, and it will be sufficient with moderate altitudes for the measurement of heights and for the reduction of barometric readings to sea level, with all the accuracy of which the problem is capable.

Explanation. In estimating heights with the barometer, the barometric readings should be reduced to Mercury at temperature 32°. The aneroid requires no correction. The temperature is the mean temperature of the air at the upper and lower station if they differ. The value in feet for *one inch* of the barometer being determined for the *mean* height of the barometer between the upper and lower stations, a simple proportion gives the value for the whole difference of readings.

It will be found very convenient to attach this little table to every aneroid.

Temperature of Mercury 32°. Temperature of Air 0°.

| Bar. in inches. | Feet due to 1 inch. | Difference per $\frac{1}{10}$ inch. |
|---|---|---|
| 30·9 | 781·2 | 3·0 |
| 30·5 | 793·0 | 3·1 |
| 30·0 | 808·3 | 3·2 |
| 29·5 | 824·1 | 3·3 |
| 29·0 | 840·6 | 3·4 |
| 28·5 | 857·8 | 3·6 |
| 28·0 | 875·8 | 3·8 |
| 27·5 | 894·7 | 4·0 |
| 27·0 | 914·5 | 4·2 |
| 26·5 | 935·4 | 4·4 |
| 26·0 | 957·4 | 4·6 |
| 25·5 | 980·5 | 4·9 |
| 25·0 | 1004·9 | |

For any temperature of the air above 0 add twice the temperature in feet.

---

* If $a$ is the total difference of level between two stations, with temperature of air 32°, for any other mean temperature T assuming expansion at $\frac{1}{490}$ per degree—

$$\text{Correction} = \frac{T - 32}{490} a.$$

# Appendix.

Example—Ben Lomond:

Bar. at base reduced to Mer. 32° or aneroid 29·803 Inches. temp. of air 59
,, at summit ,, ,, 26·606 ,, 48

Mean 28·205    Mean 53·5

Difference 3·197

Table gives for Bar. 28·205 one inch = 868 Feet. temp. of air 0
Add for temperature 53·5    107

One inch = 975 temp. of air 53·5

And 3·197 inches × 975 = 3117 feet, the height of Ben Lomond.

### CORRECTIONS OF BAROMETER.

Whether for comparison with the aneroid or with other barometers, the reading of the mercurial barometer must always be reduced to its reading for mercury at 32°.

If the temperature of the mercury is above 32°, multiply the excess by ·003 and deduct the result from the reading.

As stations are usually above the level of the sea, it is also necessary for comparison to reduce the readings to their equivalent at sea level by adding $\frac{1}{10}$ inch for every 100 feet of elevation.

For greater accuracy ascertain approximately as above the *mean* reading of the barometer between the station and the sea level, and take out the correct value due to 1 inch for the given *mean* reading and mean temperature of the air from the table at p. 353.

Example:—

Temperature of air 60°, temperature of the mercury 65°: barometer 29·89 inches.

# Barometric Heights and Corrections. 355

First, for the reading for mercury at 32° we have

33 × ·003 = ·099 and 29·89 — ·099 = 29·791 inches for mercury 32°; which will correspond with the aneroid reading.

Secondly, elevation being 200 feet. To reduce to sea level 29·791 + $\frac{2}{10}$ inch = 29·991; which is the approximate reading at sea level for mercury—32.°

More accurately, the *mean* reading between sea level and the station will be 29·791 + $\frac{1}{10}$ inch = 29·891 inch. The value in feet due to one inch for barometer 29·891 and thermometer 60°, will from the table be 815 + 120 feet, or 931 feet. Hence,

As 931 : 200 :: 1 inch : ·215 inches.

And 29·791 + ·215 = 30·006,—the correct reading at sea leve for mercury at 32°.

NOTE.—Our formula gives for the reading at sea level:—

$$Log.\ H = log.\ h + \frac{\text{Altitude}}{60158 + \frac{60158\ (2\ t\ -\ 64)}{980}}$$

FINIS.

PUBLISHED ANNUALLY.

A Very Valuable Book of Reference and a Handsome Table-Book.

### UNDER THE PATRONAGE OF

Her Majesty the  H.R.H. the Prince
Queen, of Wales,

THE HOUSE OF LORDS, THE HOUSE OF COMMONS, AND OTHER GOVERNMENT OFFICES.

LARGE SIZE PAPER, OR ROYAL EDITION, OF

## DEBRETT'S
## *Peerage, Baronetage, Knightage, and Titles of Courtesy.*

Edited by R. H. MAIR, LL.D. 166th year of Publication, greatly amplified and with improved reference arrangements. Beautifully printed, with wide margins. 1400 pp. 1500 Heraldic Engravings. Emblematical binding, cloth gilt, gilt edges, 25/-. Persian calf, full gilt edges, 30/-. Published in January.

The Lord Chancellor (Baron Cairns) thus complimented "Debrett" in his Speech in the House of Lords, on April 3rd, 1876:—

"A Depository of Information which I never open without amazement and admiration."

DEBRETT, the oldest, best patronised, most reliable and most practical Work of its kind, owes its popularity to the fact, in the aggregate, it furnishes *ten times* more information respecting living members of the Nobility than all other kindred books combined. The details it furnishes are collated from all parts of the world at an enormous cost, and no expense is otherwise spared to attain correctness. The new arrangement is unique, and entirely obviates the necessity for reference to various sections; as under the heading of each title full particulars appear respecting all living persons who are nearly related to the Peer or Baronet.

IT IS THE ONLY VOLUME THAT ATTEMPTS TO SUPPLY:—

1.—Full particulars of the Services of Peers, Baronets, Privy Councillors and Knights. 2.—Full particulars of Sons of Peers and Baronets. 3.—Addresses and Clubs of Peers' and Baronets' Sons. 5.—Addresses of Peers' and Baronets' Daughters, and of Widows of Peers, Baronets, and Knights. 5.—Particulars of all peerages that have become Extinct, Dormant or Abeyant during the present century. 6.—Complete Indices to the Names of all Sons and Daughters of Peers, and of Married Daughters of Baronets. 7.—A perfect Alphabetical Index in body of Work to all Inferior Titles borne by Peers and to surnames of Peers. 8.—Biographical Sketches of Predecessors of present Peers. 9.—Information concerning living female representives of Extinct Baronetcies.

DEAN & SON, PUBLISHERS, 160A, FLEET STREET, E.C.

DEBRETT'S HANDBOOKS OF THE ARISTOCRACY.

# CONTENTS.

Biographies of Peers (Spiritual and Temporal), Peeresses, Baronets and Knights; Widows of Peers, Baronets, Knights, and Sons of Peers; Sons and Daughters of Peers and Baronets; Privy Councillors; Protestant Bishops of Ireland and Scotland.
Index to Sons and Daughters of Peers; and to Married Daughters of Baronets.
Church Patronage.
Addresses and Clubs of persons referred to in the Work.
Surnames of Peers arranged in body of Work.
Inferior Titles of Peers, and Titles borne by eldest sons of Dukes, Marquesses and Earls (alphabetically arranged).
Extinct, Dormant, and Abeyant Peerages.
Living Female Representatives of every Extinct Peerage and

Baronetcy, arranged in body of Work under heading of Title.
Table of Precedence.
Modes of Addressing Titled Persons
Essay on Titles, Orders and Dignities
Addresses of Metropolitan and Provincial Club Houses.
Lords Lieutenant of Counties.
Chaplains to the Queen.
Royal Households.
Roman Catholic Peers and Baronets
Obituary for past year.
Peers who are Minors and Peeresses in their Own Right.
Historical and Genealogical Notes of Baronets' families.
Peers and Peeresses entitled to Quarter the Royal Arms of Plantagenet.
Complete descriptions of Armorial Bearings borne by all Peers and Baronets.
Engravings of the Heraldic Insignia of all Peers and Baronets, &c.

*The above Volume can also be had in two separate Books,—
as the Library Edition,—viz:*

**DEBRETT'S ILLUSTRATED PEERAGE AND TITLES OF COURTESY.** Published annually (January). 760 pp., cloth gilt, 13/-; half-calf and gilt edges, 17/6.

**DEBRETT'S ILLUSTRATED BARONETAGE, WITH THE KNIGHTAGE.** Published annually (January). 660 pp., cloth gilt, 13/-; half-calf and gilt edges, 17/6.

*DEBRETT'S HERALDIC AND BIOGRAPHICAL*

## House of Commons and the Judicial Bench.

1000 Coats of Arms. Cloth gilt, 7/-; half-bound calf, gilt edges, 10/6.

Contains—Biographies of Members of Parliament, with their Political Views,—Arms fully emblazoned,—Issue,—Residences and Clubs,—Church Patronage, &c. Biographies of Judges of England and Ireland,—Scottish Lords of Session,—Commissioners of Bankruptcy,—Recorders,—Judges of County Courts, &c. Counties, Cities, and Boroughs Returning Members, with Names of Representatives, Coats of Arms, &c., &c,

DEBRETT'S HANDBOOKS OF THE ARISTOCRACY ARE
THE CHEAPEST AND MOST RELIABLE WORKS OF THE KIND.

DEAN & SON, PUBLISHERS, 160A, FLEET STREET, E.C.

*Just Ready, Price Five Shillings, handsomely bound, gilt edges,*

# LEAVES FROM MY NOTE BOOK;

BEING A

*COLLECTION OF TALES, ALL POSITIVE FACTS,*

PORTRAYING

## IRISH LIFE AND CHARACTER.

By an EX-OFFICER of the ROYAL IRISH CONSTABULARY.

## CONTENTS.

**CHAPTER I.**

A Story of what happened on a Christmas Day—The Attack on Mr. Forrester's House—The Attack on the Mail Coach.

**CHAPTER II.**

Canal Passenger Boat — Athlone — The Old Bridge — Priest's Déjeuné — Captain Courtney's Wooing — Parsonetown — Faction Fight—Religious Warfare.

**CHAPTER III.**

City of Kilkenny — Reminiscences of Literary Celebrities—Old Theatre—Social Intercourse—Royal Irish Constabulary — Single Combat between a chief Constable and the Leader of a Faction—Murder in the City—Murder of the Brothers Marum in County ——.

**CHAPTER IV.**

Dingle—Fishing at Mount Eagle—Tithe Warfare—Year 1848— Her Majesty's War Steamer in Harbour—Serious Engagement between Fair Inhabitants of Dingle and Officers and Crew of Steamer — Happy Ending—The Kerry Dragoons.

**CHAPTER V.**

Cahirciven—Valentina State Quarries—Knight of Kerry—Waterville Lake, Trout Fishing—Ancient Burying Place—Story of a "Furriner" as told by Old Shawn the Boatman — Funeral Crossing the Lake — O'Connell's Birthplace.

**CHAPTER VI.**

Murder of McDermott — Difference in Religious Matters—Live, and Let Live—Sectarian Hatred—The Flitting —A New Home—The "Banshee"—Prognostics Fulfilled—Return to the Old Home, and Early Associations.

**CHAPTER VII.**

Murder of Mr. Hall—A Day's Fishing—Meeting in the Glen—Irish Hospitality—A Pleasant Evening — The First and Last Meeting.

**CHAPTER VIII.**

Battle of the Churchyard—Reformists in the West—Guarding a Priest to Church—Snakes in the Grass—A Sail in the Rector's Yacht across the Bay in company with the Priest — Narrow Escape — Hawser Cut — Vessel Run Ashore—Death of a Convert, and what followed.

**CHAPTER IX.**

Duel between Mr. Shaw, Sheriff of the Queen's County, and Mr. Cooke—Death of Mr. Shaw—Use and Abuse of Duelling—Tempora Mutantur, etc., are the changes for the better?—Fistic Duel between Dan Donelly and Oliver on English Ground; Defeat of the Latter.

**CHAPTER X.**

Duel between Captain Smith and Mr. O'Grady at Harold's Cross, Dublin—Death of O'Grady— Sentence on Smith and his Second, Lieutenant Markham.

**CHAPTER XI.**

Duel between Captain Smith, 59th Regiment (Uncle to Smith who shot O'Grady in Dublin) and Colonel Macdonald, 92nd Highlanders, at Fermoy—Death of Captain Smith—Duelling *versus* Divorce Court.

**CHAPTER XII.**

Torc Lake—The O'Donoghue—The Echoes of Killarney—Paddy Blake's Echo- Mangerton.

DEAN & SON, PUBLISHERS, 160A, FLEET STREET E.C.

# Well Selected Presentation Books

### FOR BOYS AND GIRLS, YOUNG MEN AND LADIES.

The following list may be relied upon as containing books of a thoroughly interesting character, and at the same time thoroughly free from vice; books, in fact, that can be read in the Family Circle. Each book is very handsomely bound in full gilt cloth, with block design on cover, bevelled boards, and gilt edges. 3s. 6d. each.

**REMARKABLE MEN:** their Lives and Adventures. By M. S. COCKAYNE. Fully Illustrated. Cloth, full gilt sides and edges, 3s. 6d.

**MEN OF DEEDS AND DARING.** The Story and Lesson of their Lives. By E. N. MARKS. Fully Illustrated. Cloth, full gilt sides and edges, 3s. 6d.

**SAYINGS, ACHIEVEMENTS, and INTERVIEWS of GREAT MEN.** By the Author of "Heroines of Our Own Time." Fully Illustrated. Cloth, gilt sides and edges, 3s. 6d.

"These stories of the Lives of Illustrious Men will be found to be prepared with so much care and made so very interesting that these three books are sure to become great favourites in every family where young people love to read; such books as these instruct, elevate, and create a noble spirit and a desire to excel in the sphere of life marked out for us."—*Oxford Times.*

**LIFE and FINDING of Dr. LIVINGSTONE;** containing Original Letters by H. M. STANLEY. Portraits and numerous Illustrations. Cloth, gilt sides and edges, 3s. 6d.

**BOOK of WONDERS, EVENTS, and DISCOVERIES.** Edited by JOHN TIMBS, Author of "Things not Generally Known," &c. Fully Illustrated. Cloth, full gilt sides and edges, 3s. 6d.

**ONE HUNDRED and FIFTY BIBLE PICTURES and STORIES.** By Mrs. UPCHER COUSENS, Author of "Pleasant Sundays," Editor of "Happy Sundays," &c. Fully Illustrated. Cloth, full gilt sides and edges, 3s. 6d.

**NOTABLE WOMEN.** By ELLEN C. CLAYTON. Illustrated. Cloth, full gilt sides and edges, 3s. 6d.

**WOMEN of the REFORMATION:** their LIVES, TRAITS, and TRIALS. By ELLEN C. CLAYTON. Fully Illustrated. Cloth, full gilt sides and edges, 3s. 6d.

**CELEBRATED WOMEN.** A Book for Young Ladies. By ELLEN C. CLAYTON. Handsomely bound, cloth, gilt edges, 3s. 6d.

**MINISTERING WOMEN.** Edited by Dr. CUMMING. Cloth, gilt sides and edges, 3s. 6d.

**Miss MILLY MOSS; or, SUNLIGHT and SHADE.** By ELLEN C. CLAYTON. Cloth, gilt sides and edges, 3s. 6d.

**FRIENDS IN FUR.** Amusing and True Tales, by Dr. STABLES. With Eight Illustrations. Price 3s. 6d. cloth gilt edges.

**ALONE; or Two Thousand Pounds Reward.** A Tale of London Life. By Mrs. E. A. MELVILLE. Frontispiece. Cloth gilt, 2s. 6d.

---

DEAN & SON, PUBLISHERS, 160A, FLEET STREET, E.C.

# Deeds of Daring Library

PEOPLE of every age, taste, and class, take delight in reading of feats of heroism, dangers bravely encountered, and perils nobly overcome. MESSRS. DEAN & SON, therefore, are now publishing a series of such narratives in a cheap and popular form, under the title of

## Deeds of Daring Library,

BY

## LIEUT.-COLONEL KNOLLYS, F.R.G.S.,

AND OTHERS.

A Complete Book is issued at intervals, each with a Frontispiece in Colours and numerous full-page Wood Engravings; price One Shilling and Sixpence, handsomely bound in cloth gilt, for prizes and presents Also bound in fancy wrappers in colours, at One Shilling each

*Among the Series are the following:—*

1—SHAW, THE LIFE GUARDSMAN. Ready.
2—THE EXPLOITS OF LORD COCHRANE. Ready.
3—THE VICTORIA CROSS IN THE CRIMEA. Ready.
4— ,, ,, INDIA. Ready.
5— ,, ,, THE COLONIES, &c. In the Press.
11—BRAVE DEEDS AND HEROIC ACTIONS. Series I.—MILITARY. By CAPTAIN CLAYTON. Ready.
12—BRAVE DEEDS AND HEROIC ACTIONS. Series II.—NAVAL AND MILITARY. By CAPTAIN CLAYTON. Ready.
13—PAUL JONES, THE NAVAL HERO OF AMERICAN INDEPENDENCE BY JAMES WARD. Ready.

*The following, by Lieut.-Major Knollys, are in active preparation:*
6—UNDECORATED MILITARY AND NAVAL HEROES.
7—GALLANT SEPOYS AND SOWARS.
8—DARING DEEDS AFLOAT—ROYAL NAVY.
9— ,, ,, ,, MERCHANT NAVY AND PRIVATEERS.
10—FEMALE HEROISM IN WAR.

DEAN & SON, PUBLISHERS, 160A, FLEET STREET, E.C.

www.ingramcontent.com/pod-product-compliance
Lightning Source LLC
Chambersburg PA
CBHW020226240426
43672CB00006B/435